Lecture Notes in Biomathematics

98

Jin Yoshimura   Colin W. Clark (Eds.)

Adaptation in Stochastic Environments

Springer-Verlag Berlin Heidelberg GmbH

Editors

Jin Yoshimura
Dept. of Biological Sciences,
Binghamton University
Binghamton NY 13902-6000
USA

Colin W. Clark
Institute of Applied Mathematics
The University of British Columbia
Vancouver, B.C.
Canada V6T 1Z2

Mathematics Subject Classification (1991): 92D15, 92D25, 92D40, 92D50,

ISBN 978-3-540-56681-6   ISBN 978-3-642-51483-8 (eBook)
DOI 10.1007/978-3-642-51483-8

Typesetting: Camera-ready by author/editor
46/3140-543210 - Printed on acid-free paper

PREFACE

The classical theory of natural selection, as developed by Fisher, Haldane, and Wright, and their followers, is in a sense a *statistical* theory. By and large the classical theory assumes that the underlying environment in which evolution transpires is both constant and stable – the theory is in this sense *deterministic*. In reality, on the other hand, nature is almost always changing and unstable.

We do not yet possess a complete theory of natural selection in stochastic environments. Perhaps it has been thought that such a theory is unimportant, or that it would be too difficult. Our own view is that the time is now ripe for the development of a probabilistic theory of natural selection. The present volume is an attempt to provide an elementary introduction to this probabilistic theory. Each author was asked to contribute a simple, basic introduction to his or her specialty, including lively discussions and speculation. We hope that the book contributes further to the understanding of the roles of "Chance and Necessity" (Monod 1971) as integrated components of adaptation in nature.

The book contains ten chapters. The first two chapters serve as an introduction: first Yoshimura and Clark discuss the historical background of probabilistic theories of adaptation and selection, and consider some of the conceptual issues and developments arising from the study of evolution in stochastic environments. Next Cohen discusses fitness optimization criteria under a variety of stochastic assumptions. The selection regime depends both on the life history details and on the patterns of environmental and demographic variability. Cohen provides a list of such selection regimes (fitness criteria) in stochastic environments.

The next two chapters describe life-history evolution in stochastic environments. Using the concept of an evolutionarily stable strategy (ESS), which takes on new dimensions in the stochastic setting, Kisdi and Meszéna analyze optimality and coexistence conditions for life-history traits under density-dependent regulation. In their models fitness is affected not only directly by environmental fluctuations, but also indirectly by fluctuating population density. Orzack investigates density-independent population dynamics with complex life-history traits in age-structured populations in stochastic environments. He particularly considers stochastic demography with neutrality (or indifference) of stochastic growth rates or extinction probabilities. Orzack also suggests that, perhaps contrary to presently received wisdom, the case for density-dependent population regulation in stochastic environments is still not scientifically convincing.

Further theoretical aspects of evolution and adaptation in stochastic environments are taken up in the next three chapters. León considers phenotypic plasticity, using Levins's fitness-set analysis. He considers especially the problem of costs associated with plasticity, as well as the possibility that information (i.e. environmental cues) is imperfect. Next, Clark and Yoshimura consider the interplay between demographic and environmental stochasticity in a behavioral context of foraging under risk of predation, using an ESS analysis. They also briefly consider a genetic version of the stochastic evolutionary game. Beissinger and Gibbs show how to quantify environmental predictability, using as an example weather records from tropical and subtropical rainforests. They compare two measures of predictability, Colwell's index and spectral densities, together with two simple statistical measures of variability, standard deviation and the coefficient of variation.

The next two chapters concern the application of stochastic modeling to problems in genetics and community ecology. Hazel and Smock discuss a quantitative genetic model of conditional strategies as threshold traits. They also discuss adaptive evolution and maintenance of conditional strategies in spatially or temporally varying environments. Sevenster and van Alphen analyze the life-history tradeoff between fast-growing, short-lived species in frugivorous Drosophila communities when food resources (fruits) are variable. They suggest that coexistence of the two life-history strategies may be maintained under certain restricted conditions.

The final chapter, by Cohen, departs from previous chapters, in passing from intergenerational stochasticity (coarse-grained environments) to consider foraging behavior of animals faced with environmental fluctuations within their life spans (fine-grained environments). Optimal foraging theory has been extensively developed and tested in recent decades; Cohen employs the ESS concept in a model of adaptive learning, considering the question of the optimal allocation of time between searching and foraging.

Most of the papers in this volume were presented at a workshop on "Adaptation in a stochastic environment," held at the Second International Congress of the European Society for Evolutionary Biology, Debrecen, Hungary, 3–4 September, 1991. The workshop was organized by Jin Yoshimura, Géza Meszéna, and Éva Kisdi, and chaired by Colin Clark. The details of organization and presentation were performed admirably by Géza and Éva, who originally invited Jin as organizer. We are indebted to them for their contributions, without which this volume would not have been possible. We also acknowledge the outstanding contributions of Dan Cohen, who was prevented from attending the workshop at the last minute by a sudden but temporary illness. His encouragement was always present.

Our special thanks to Janet Clark, who reformatted and TEX'ed the manuscripts into a uniform typographical style for the volume. We also wish to thank many colleagues who served as reviewers for the various chapters. Finally, we thank all the contributors for their patience and forbearing – not to mention their outstanding contributions to the volume. CWC has been partly supported by NSERC grant 83990. During the editorial work JY was supported by U.S. Department of Energy grant DE-FG02-89ER60884 to David Sloan Wilson, and PHS grant GM37841 to Marcy K. Uyenoyama.

Jin Yoshimura, Durham, NC
Colin W. Clark, Vancouver, B.C.
January 17, 1993

TABLE OF CONTENTS

Contents

INTRODUCTION: HISTORICAL REMARKS

Jin Yoshimura
Department of Zoology
Duke University
Durham, NC 27708-0325
and
Colin W. Clark
Institute of Applied Mathematics
The University of British Columbia
Vancouver, B.C. V6T 1Z2

What is maximized by natural selection? A general answer is the expected reproductive success of the individual; i.e. mean individual fitness. This basic fitness maximization principle underlies much of the modern evolutionary theory of adaptation (Fisher 1930, Williams 1966, Maynard Smith 1978). Criticism of the hypothesis of fitness maximization has, however, also been widespread (e.g. Dupré 1987).

If the environment is stochastic, that is, changing and uncertain, this general answer – a commonly held view of adaptation – is no longer adequate or correct. In a stochastic environment, mean fitness is not the sole indicator of the force of natural selection; variance, or more generally the entire probability distribution of fitness, becomes a critical factor of selection (Yoshimura and Shields 1987). Geometric mean fitness is a typical example of such a selection criterion under environmental stochasticity over many generations (Lewontin and Cohen 1969, Yoshimura and Clark 1991).

Nature is almost always stochastic. By definition no two organisms can share an identical environment; the experiences of individuals are always different. Physical and biotic environments are continuously varying both spatially and temporally. The inevitable environmental stochasticity encountered by individual organisms implies that conceptualizing adaptive evolution in terms of mean fitness – a natural extension of deterministic theories of natural selection – is likely to be biased and misleading.

Our understanding of stochastic selection is still immature, however. Most current stochastic models are concerned with adaptive responses to cope with environmental stochasticity, such as phenotypic plasticity (Via and Lande 1985), or bet-hedging life-history traits such as iteroparity or dispersal (Seger and Brockmann 1987, Philippi and Seger 1989), and adaptive behavioral responses (Mangel and Clark 1988), including learning (Oaten 1977).

The optimization principle of natural selection becomes different in a stochastic framework, as we have already mentioned. The general view of adaptation in a stochastic environment is therefore fundamentally different from that within a deterministic setting (Yoshimura and Clark 1991). The present volume focuses on this general stochastic, or probabilistic, theory of natural selection. The various authors present their own perspectives, aiming at this general view of adaptation in a stochastic environment.

Early studies on uncertainty

The earliest stochastic models of natural selection, which go back to the 1950s, were genetic models dealing with the conditions under which genetic polymorphism and heterogeneity are maintained (Levene 1953, Dempster 1955, Li 1955, Haldane and Jayakar 1963;, see also Cannings 1971, Roughgarden 1979). Levene (1953) modeled spatial heterogeneity as a special form of niche differentiation, where each niche is limited in resources; later his model was shown to use the geometric mean maximization principle (Li 1955, Cannings 1971). Subsequently the geometric mean concept of fitness was used to consider temporal heterogeneity across generations (Dempster 1955, Haldane and Jayakar 1963). These models assumed that the fitness of a genotype was dependent on the environmental state.

Beginning in the 1960's a long series of ecological models was developed. First of all, Dan Cohen, a contributor to this volume, developed a series of elegant stochastic optimization models (Cohen 1966, 1967, 1968, 1970, 1976). The main advantage of these ecological models is that the stochasticity of environments is fully incorporated into fairly complex life history strategies, rather than merely determining the fitness of a given genotype. Their major drawback, obviously, is the lack of genetic components.

The analysis of geometric mean fitness in both genetic and ecological models led to important statistical properties of population growth in stochastic settings (Lewontin and Cohen 1969, Slatkin 1974, Gillespie 1977). These analyses indicated the nature of variance discounting in a stochastic environment. For example, Verner (1965) described the nature of variance discounting in a model of sex-ratio evolution.

In the meantime, Levins (1968) developed an elegant graphical method – the fitness set – to analyze phenotypic adaptation in changing environments. He assumed that the fitness of a given phenotype was a function of environmental conditions; this work bridges the gap between genetic and ecological models. Although the graphical method applies only to simple cases, it is a very powerful and illustrative technique for treating optimization principles in stochastic environments. Levins' analysis also characterizes the difference between arithmetic mean and geometric mean fitnesses, and shows how these concepts apply to fine-grained or coarse-grained environments, respectively.

Another problem involving stochasticity or uncertainty in phenotypic adaptation arises from the fact that phenotypes are the product of both a genotype and an environment. Consequently phenotypic expression may have an inevitable variance. With such phenotypic variation, genetic optimality usually differs from phenotypic optimality unless there is symmetry in both the fitness potential and the distribution of phenotypes (Yoshimura and Shields 1987). Mountford (1968) discussed this discrepancy in phenotypic and genetic optimality for the case of litter (clutch) size.

Several developments in stochastic life history theory also occurred in the late 1960s. Den Boer (1968) developed a notion of "risk spreading," considering a number of life history phenomena as adaptations to environmental stochasticity, including phenotypic variation, dispersal, and dormancy (see Root and Kareiva 1984). Murphy (1968) and Schaffer (1974) considered iteroparity as an adaptation to environmental stochasticity. Stearns (1976) reviewed and discussed extensively many aspects of life history adaptations to stochasticity.

Stochastic optimization principles

The most profound finding in the study of adaptation in stochastic environments concerns the nature of optimization principles, or criteria, in the stochastic setting (Yoshimura and Shields 1992). As noted above, the standard arithmetic mean fitness (expected per capita population growth rate) is replaced by the geometric mean fitness under intergenerational stochasticity, as well as in certain forms of spatial heterogeneity.

Phenotypic variance is another example. Under stochastic phenotypic variation, the genetic optimum is given by a weighted average of phenotypic fitness (Mountford 1968, Yoshimura and Shields 1987). This shift in optimization principles shows that the standard arithmetic mean theory of natural selection is invalid in principle. In the two examples mentioned here (fitness and phenotypic variance), optimization must be assessed at the population level for a given genetic strategy, and not at the individual level (Yoshimura and Clark 1991).

Stochastic fluctuations are also the source of individual behavioral variation. Within a generation any individual organism faces environmental unpredictability in its daily life. Patchiness of a forager's environment, for example, has been the subject of many studies, both theoretical and empirical, in behavioral ecology (MacArthur and Pianka 1966, Charnov 1976, Stephens and Krebs 1986). The standard optimization criterion used in early work in this area has been the long-term average rate of net energy gain. As shown by Caraco, however, foragers often display risk sensitivity, treating high and low variance options differently (see Caraco et al. 1980). In this respect, the maximization of expected utility (a concept related to the geometric mean) is a better predictor of foraging behavior than is the maximization of average net energy gain (see Real 1980, Stephens 1981, Real and Caraco 1986, Stephens and Krebs 1986).

A fully stochastic, dynamic state-variable approach to the modeling of behavioral decisions of all kinds has been developed by Houston and McNamara (1988) and Mangel and Clark (1988). This theory is fully adequate for fine-grained environments, but encounters severe difficulties of implementation in the case of coarse-grained environments (Yoshimura and Clark 1991).

Another important issue associated with environmental stochasticity is the possibility of extinction (Harrison and Quinn 1991). Geometric mean fitness characterizes long-term population growth, but cannot account directly for the probability of extinction. Rather, the geometric mean likely correlates with the extinction probability (Lewontin and Cohen 1969). Extinction is obviously a critical problem for genetic strategies; some authors actually characterize the probability of extinction as the fundamental measure of fitness (e.g. Cooper 1984). However, the balance between local population extinction

and reintroduction of strategies via mutation and migration also has to be considered
(e.g. Lande 1987, Schoener and Spiller 1987, McCauley 1989). The characterization of
extinction probabilities is obviously a crucial issue in conservation biology (Lande 1988,
Brussard 1991).

Adaptive strategies and stochastic evolutionary games

In a stochastic environment the concept of an evolutionary game also becomes quite
different from that in a stable, or deterministic environment. The concept of an evolutionarily stable strategy (ESS) as a strategy that *cannot* be invaded by alternative
strategies, does not apply in the stochastic setting (see Levin et al. 1984, Ellner 1985).
Both environmental and demographic variation may cause tradeoffs in phenotypic traits,
yielding fluctuating abundances of two or more almost equally adaptive strategy morphs
(Yoshimura and Clark 1991). In this context, environmental stochasticity may also promote the coexistence of species in community dynamics (Chesson and Warner 1981,
Chesson 1985). Both evolutionary stability and adaptation become only relative, probabilistic terms with which to compare different strategies. No single strategy may be
uniquely adaptive, or evolutionarily stable; instead, several different strategies may be
evolutionarily compatible (Cohen and Levin 1991, Yoshimura and Clark 1991).

Adaptation to stochasticity

Two distinct types of strategy exist to cope with stochastic environments: risk avoidance
and risk spreading (den Boer 1968, Seger and Brockmann 1987). Risk avoidance is an
individual adaptation associated with the reduction in the variance of fitness, and as a
result, an increase in geometric mean fitness.

Risk spreading strategies, on the other hand, are adaptations at the population
level (although individuals may apply risk spreading diversification to their offspring),
whereby individual members are spread or diversified into different habitats, times, or
strategies. Temporal diversification is achieved by various life history traits, such as
iteroparity or asynchronous breeding (Tuljapurkar and Orzack 1980, Tuljapurkar 1989),
and seed dormancy (Cohen 1970, 1976). Spatial diversification is achieved by dispersal
of offspring, whether uniformly or differently between individuals (Root and Kareiva
1984). Temporal and spatial diversification may also be combined (Levin et al. 1984).
Strategy diversification is called polyphenism, or phenotype-limited polymorphism, and
includes so-called adaptive coin flipping (Cooper and Kaplan 1982). It is important to
note that under optimal risk spreading, individuals need not have identical fitness; some
options may have almost zero chance of success except under special, rare circumstances
(Yoshimura and Clark 1991).

The extent to which an organism is able to sense changes in its environment, and
respond accordingly, is an important consideration in the study of adaptation under uncertainty. Phenotypic and behavioral plasticity refer to such reactions to environmental
conditions. In some cases organisms may be able to make short-term predictions of
future conditions, using various cues, and to respond in an anticipatory manner (Clark,

submitted). Learning, that is, the assessment of environmental states, is itself an important component of adaptive behavior (Oaten 1977, Stephens 1989, Mangel and Clark 1988).

Information about a stochastic system is never complete. An organism can only respond to the information that it currently possesses; the possibility that this information may be misleading must be allowed for, if it affects fitness. It is the very nature of uncertainty that "optimal" adaptations must make the best of a bad job.

Conclusions

We have tried to highlight the major themes of, and early contributions to, the theory of adaptation in stochastic environments. As the reader of this volume will soon learn, a great deal of water has passed over the dam since the publication of these articles. Concepts have been extended, tightened, and refined. Simple early models have blossomed into general theories. New mathematical techniques have been developed, and old ones have found new application. Although much interesting and important work remains to be done, we can now begin to see more clearly how chance and necessity, to use Monod's (1971) perceptive phrase, are essential, creative forces in evolution.

References

Brussard, P.F. 1991. The role of ecology in biological conservation, *Ecological Applications* 1: 6–12.

Cannings, C. 1971. Natural selection at a multiallelic autosomal locus with multiple niches, *J. Genet.* 60: 255–259.

Caraco, T., S. Martindale, and T.S. Whittam. 1980. An empirical demonstration of risk-sensitive foraging preferences, *Anim. Behav.* 28: 820–830.

Charnov, E.L. 1976. Optimal foraging: the marginal value theorem, *Theoret. Pop. Biol.* 9: 129–136.

Chesson, P.L. 1985. Coexistence of competitors in spatially and temporally varying environments: a look at the combined effects of different sorts of variability, *Theoret. Pop. Biol.* 28: 263–287.

Chesson, P.L. and R.R. Warner. 1981. Environmental variability promotes coexistence in lottery competitive systems, *Amer. Nat.* 117: 923–943.

Cohen, D. 1966. Optimizing reproduction in a randomly varying environment, *J. Theoret. Biol.* 12: 119–129.

Cohen, D. 1967. Optimizing reproduction in a randomly varying environment when a correlation may exist between the conditions at the time a choice has to be made and the subsequent outcomes, *J. Theoret. Biol.* 16: 1–14.

Cohen, D. 1968. A general model of optimal reproduction in a randomly varying environment, *J. Ecol.* 56: 219–228.

Cohen, D. 1970. A theoretical model for the optimal timing of diapause, *Amer. Nat.* 104: 389–400.

Cohen, D. 1976. The optimal timing of reproduction, *Amer. Nat.* 110: 801–807.

Cohen, D., and S.A. Levin. 1991. Dispersal in patchy environments: the effects of temporal and spatial structure, *Theoret. Pop. Biol.* 39: 63–99.

Cooper, W.S. 1984. Expected time to extinction and the concept of fundamental fitness, *J. Theoret. Biol.* 107: 603–629.

Cooper, W.S. and R.H. Kaplan. 1982. Adaptive "coin-flipping": a decision-theoretic examination of natural selection for random individual variation, *J. Theoret. Biol.* 94: 135–151.

Dempster, E.R. 1955. Maintenance of genetic heterogeneity, Cold Spring Harbor Symp. *Quant. Biol.* 20: 25–32.

den Boer, P.J. 1968. Spreading of risk and stabilization of animal numbers, *Acta Biotheoretica* 18: 165–194.

Dupré, J. (ed.). 1987. *The Latest on the Best: Essays in Evolution and Optimality.* MIT Press, Cambridge, MA.

Ellner, S. 1985. ESS germination strategies in randomly varying environments. I. Logistic-type models, *Theoret. Pop. Biol.* 28: 50–79.

Fisher, R.A. 1930. *The Genetical Theory of Natural Selection.* Clarendon Press, Oxford.

Gillespie, J.H. 1977. Natural selection for variance in offspring numbers: a new evolutionary principle, *Amer. Nat.* 111: 1010–1014.

Haldane, J.B.S. and S.D. Jayakar. 1963. Polymorphism due to selection of varying direction, *J. Genet.* 58: 237–242.

Harrison, S. and J.F. Quinn. 1989. Correlated environments and the persistence of metapopulations, *Oikos* 56: 293–298.

Houston, A.I. and J.M. McNamara. 1988. A framework for the functional analysis of behavior, *Behav. Brain Sci.* 11: 117–163.

Lande, R. 1987. Extinction thresholds in demographic models of territorial populations, *Amer. Nat.* 130: 624–635.

Lande, R. 1988. Genetics and demography in biological conservation, *Science* 241: 1455–1460.

Levene, H. 1953. Genetic equilibrium when more than one niche is available, *Amer. Nat.* 87: 331–333.

Levin, S.A., A. Hastings and D. Cohen. 1984. Dispersal strategies in patchy environments, *Theoret. Pop. Biol.* 26: 165–191.

Levins, R. 1968. *Evolution in Changing Environments.* Princeton University Press, Princeton, N.J.

Lewontin, R.C. and D. Cohen. 1969. On population growth in a randomly varying environment, *Proc. Nat. Acad. Sci. USA* 62: 1056–1060.

Li, C.C. 1955. The stability of an equilibrium and the average fitness of a population, *Amer. Nat.* 89: 281–295.

Mangel, M. and C.W. Clark. 1988. *Dynamic Modeling in Behavioral Ecology.* Princeton University Press, Princeton, NJ.

Maynard Smith, J. 1978. Optimization theory in evolution, *Ann. Rev. Ecol. Syst.* 9: 31–56.

MacArthur, R.H. and E.R. Pianka. 1966. On the optimal use of a patchy environment, *Amer. Nat.* 100: 603–609.

McCauley, D.E. 1989. Extinction, colonization, and population structure: a study of the milkweed beetle, *Amer. Nat.* 134: 365–376.

Monod, J. 1971. *Chance and necessity*. Vintage Books, New York.

Mountford, M.D. 1968. The significance of litter-size, *J. Anim. Ecol.* 37: 363–367.

Murphy, G.I. 1968. Pattern in life history and the environment, *Amer. Nat.* 102: 390–403.

Oaten, A. 1977. Optimal foraging in patches: a case for stochasticity, *Theoret. Pop. Biol.* 12: 263–285.

Philippi, T. and J. Seger. 1989. Hedging one's evolutionary bets, revisited, *Trends in Evolutionary Ecology* 4: 41–44.

Real, L. 1980. Fitness, uncertainty, and the role of diversification in evolution and behavior, *Amer. Nat.* 115: 623–638.

Real, L. and T. Caraco. 1986. Risk and foraging in stochastic environments, *Ann. Rev. Ecol. Syst.* 17: 371–90.

Root, R.B. and P.M. Kareiva. 1984. The search for resources by cabbage butterflies (*Pieris rapae*): Ecological consequences and adaptive significance of Markovian movements in a patchy environment, *Ecology* 65: 147–165.

Roughgarden, J. 1979. *Theory of Population Genetics and Evolutionary Ecology: an Introduciton*. MacMillan, New York.

Schaffer, W.M. 1974. Optimal reproductive effort in fluctuating environments, *Amer. Nat.* 108: 783–790.

Schoener, T.W. and D.A. Spiller. 1987. High population persistence in a system with high turnover, *Science* 330: 470–477.

Seger, J. and J. Brockmann. 1987. What is bet-hedging? *Oxford Surveys in Evolutionary Biology*. 4: 182–411.

Slatkin, M. 1974. Hedging one's evolutionary bets, *Nature* 250: 704–705.

Stearns, C.S. 1976. Life history tactics: A review of the ideas, *Quart. Rev. Biol.* 51: 3–47.

Stephens, D.W. 1981. The logic of risk sensitive foraging preferences, *Anim. Behav.* 29: 628–629.

Stephens, D.W. 1989. Variance and the value of information, *Amer. Nat.* 134: 128–140.

Stephens, D.W. and J.R. Krebs. 1986. *Foraging Theory*. Princeton Univ. Press, Princeton, NJ.

Tuljapurkar, S.D. 1989. An uncertain life: demography in random environments, *Theoret. Pop. Biol.* 35: 227–294.

Tuljapurkar, S.D. and S.H. Orzack. 1980. Population dynamics in variable environments I. Long-run growth rates and extinction, *Theoret. Pop. Biol.* 18: 314–342.

Verner, J. 1965. Selection for the sex ratio, *Amer. Nat.* 99: 419–421.

Via, S. and R. Lande. 1985. Genotype-environment interaction and the evolution of phenotypic plasticity, *Evolution* 39: 505–522.

Williams, G.C. 1966. *Adaptation and Natural Selection*. Princeton Univ. Press, Princeton, NJ.

Yoshimura, J. and C.W. Clark. 1991. Individual adaptations in stochastic environments, *Evol. Ecol.* 5: 173–192.

Yoshimura, J. and W.M. Shields. 1987. Probabilistic optimization of phenotype distributions: a general solution for the effects of uncertainty on natural selection? *Evol. Ecol.* 1: 125–138.

Yoshimura, J. and W.M. Shields. 1992. Components of uncertainty in clutch-size optimization, *Bull. Math. Biol.* 54: 445–464.

FITNESS IN RANDOM ENVIRONMENTS

Dan Cohen

Department of Evolution, Systematics and Ecology
The Hebrew University
Jerusalem 91904, Israel

Abstract. The effect of random environmental changes on fitness is modelled for a wide range of life histories and patterns of environmental and demographic variability with the following main conclusions:

1. Independent individual variation in fitness does not influence selection on an allele or a trait in a large mixed population.
2. Mean log fitness is the measure of long term fitness when fitness varies between nonoverlapping generations. In this case there is no coexistence between alternative strategies.
3. With long-lived overlapping generations, a mean ratio greater than 1 between the fitness of a rare type and the fitness of the common type is the criterion for the increase of the rare type. This gives a selective advantage to the rare type, which allows coexistence between strategies that have some independently varying fitness. This also applies to a long lived seed bank in the soil.
4. In very small populations, *especially with single founders*, the long term fitness, which is given by the probability of establishment, is determined by the probability of extinction, which increases when the variance of individual fitness increases for the same mean fitness. Mutants with higher mean fitness have a higher long term fitness however, and eventually displace low variance mutants established earlier. A long term selective advantage for low variance mutants is maintained in continuously changing environment, or in structured populations whose subpopulations periodically go extinct.
5. The variation of different components of fitness does not influence fitness if it occurs independently in different individuals in large populations. However, it does influence fitness if it is synchronised to some extent in the whole population, for example in synchronised age dependent cohorts in nonoverlapping generations.

6. The optimal tradeoff between reproduction and survival is expected to take into account only the mean survival if survival is distributed independently in a large population, e.g. risk of predation, while the variance of survival has to be taken into account if mortality is synchronised, or when there is a single or few founders.

7. The variation of resources which contribute to individual fitness, e.g. food, decreases fitness over the convex range of the fitness function and increases fitness over the concave range.

8. Behavioural, developmental, or life history strategies which reduce the variance between generations are selected for: e.g. dormancy, long living perennial habit in plants, dispersal, and phenotypic polymorphism.

Introduction

Life history strategies are programs of decision rules where an individual chooses between available developmental or behavioural alternatives at critical control points. Natural selection favours those strategies which produce the highest lifetime fitness in a given environment. Modelling the optimal fitness of organisms provides relevant testable predictions about the expected natural patterns of life histories of organisms in different environments (Stearns 1976, 1992).

In most natural environments, any decision at a control point in the life of an organism leads to a random set of potential outcomes. The probabily distribution of these outcomes is determined by random environmental fluctuations or heterogeneity, and/or by random encounters with individuals of the same or of other species. The randomness and unpredictability of natural environments has important consequences for the evolution of life histories and behavioural strategies of animals and plants (Frank and Slatkin 1990, Lacey et al. 1983, Real 1980, Rubenstein 1982, Slatkin 1974, Yoshimura and Clark 1991, Real and Ellner 1992, Stearns 1976, 1992, (Kisdi and Meszéna, Orzack, and Clark and Yoshimura in this volume)).

The fitness associated with any particular life history strategy must be defined therefore by averaging the outcomes at any such decision point in a proper manner.

The question which we address in this paper is therefore:

What is the appropriate averaging function to measure fitness? i.e. "What exactly does long term natural selection select for?" We define fitness operationally in the Darwinian sense as that measure which is maximised by long term natural selection. A stricter definition is that of an ESS: An evolutionary stable strategy is stable against invasion by any rare mutant strategy (Maynard Smith 1982).

The problem has been dealt with by a number of investigators. Gillespie (1977) has modelled the effect of individual variance of fitness on selection. Frank and Slatkin (1990) and Yoshimura and Clark (1991, and in this volume), have developed very general models for selection in varying environments. Tuljapurkar (1989) has modelled demographic processes of age dependent populations in varying environments.

In this paper I construct and analyse several models which deal with the effect of varying selection in important representative types of life histories and environments, which cover a wide range of natural situations. For each model I derive the mathematical expressions for the appropriate definitions of long term fitness and of the conditions for fitness equilibrium or coexistence under varying selection.

I present each model in its most simple and general form, which can therefore usually be applied to a wide range of more specific environments and organisms that share the same general form. In this way it is possible to identify and analyse the essential characteristics of the effects of varying fitness over a wide range of patterns of variation between individuals of different life history and behaviour, and over a range of representative temporal and spatial scales of variation of the environment.

The models

This paper will mainly deal with the life histories of plants in a seasonal environment, but the models will remain general. Most of the results of the models can be generalised therefore to other types of organisms and environments, and I shall deal with a range of plant types and environmental variations.

I introduce an additional simplification to the models by assuming an asexual mode of inheritance, as if asexual clones are competing in any particular environment. This approach is usually a good approximation for selection on quantitative characters, and when genetic relatedness and family and social structures can be ignored (Maynard Smith 1982).

I. Annual plants with nonoverlapping generations; A large population

No seed dormany. All the seeds that are produced at the end of one growing season germinate at the beginning of the next growing season. All the seeds are uniformly dispersed over an area which is large relative to the distance between the individual plants, or the size of the environmental patches.

This model represents also a wide range of other organisms with nonoverlapping generations, e.g. annual species of snails, insects, lizards, etc. which reproduce once in their life time.

Random variations in longevity and seed yield between individual plants in the same year. The distribution of the random events is identical in all years, i.e. the environment may be considered constant between years.

In this case, the fitness W is the expectation or arithmetic mean of the annual seed yield, averaged over all the plants in the population in the same year:

$$W_i = Y_i = E(N_i y_i) \tag{1}$$

where W is fitness, Y is the mean annual per capita seed yield, N represents numbers or densities, y represents the per capita life time seed yield of individual plants, i represents a particular strategy, and the expectation E represents an averaging operation of the individual seed yields over all the plants.

Proof. The fraction of any one strategy in the seed population at the end of any one season is the fraction of that strategy in the previous season multiplied by the ratio between the mean per capita annual seed yield of that strategy and the mean per capita seed yield in the whole population. This is given in Equation (2):

$$N_i/\widehat{N}(t+1) = N_i/\widehat{N}(t) \times Y_i/\widehat{Y} \tag{2}$$

where $\widehat{}$ represents the mean of all the strategies in the population.

The population will be dominated eventually by the strategy which maximises the mean annual per capita seed yield, Y. If fitness is frequency dependent, the ESS will be the strategy or a set of strategies which have a local maximum of Y.

Note that the variance in the yields of the individual plants in the same year does not effect the fitness of a strategy. Optimal strategies may therefore have a very high variance of the yield of the individual plants. In fact, extremely high variance in the survival and fecundity of individuals in a population is typical of most high-fecundity highly successful species of annual plants and animals.

Random variation between years. The mean annual seed yield in the population varies randomly between years according to some given distribution.

In this case, long term fitness over a large number of years is the geometric mean or the equivalent mean logarithm of the mean annual seed yield, i.e:

$$W = E(\log Y_t) \tag{3}$$

where Y_t is the mean annual per capita seed yield at year t.

The Proof has been derived by Lewontin and Cohen (1969). Qualitatively, it is easy to see that the long term growth of any strategy over a number of years is the multiplicative series of the mean annual per capita seed yields. Such a series can be represented by the geometric mean or by the equivalent mean logarithm of the mean annual per capita seed yield Y_i.

The condition for the ESS of a strategy Y^* is therefore:

$$E(\log(Y_i/Y^*)) < 0 \tag{4}$$

or equivalently:

$$E(\log Y_i) < E(\log Y^*)$$

for any strategy Y_i, *independent of any other property of the distributions of Y_i and Y^**. In particular, according to this model, the condition for the ESS is independent of any correlations between the distributions, and no polymorphism or coexistence of more than one strategy is possible. Rather, there is always a single monomorphic ESS Y^* which maximizes $E(\log Y)$.

Using a Taylor's expansion around $E(Y)$, and neglecting higher terms, we get:

$$E(\log Y) \simeq \log E(Y) - \frac{1}{2} V(Y)/E(Y)^2 \tag{5}$$

with V indicating the variance. Following Gillespie (1977), we see therefore that the logarithmic fitness of a strategy decreases as a function of the variance of the annual yield. In particular, $E(\log Y)$ is very sensitive to rare low levels of Y.

Annual plants with delayed germination. This model represents also many cases of annual animals with dormant eggs or other resistant forms, in which only a fraction of the eggs hatch during subsequent growing seasons. For example: dormant eggs of some aquatic insects and crustaceae that inhabit temporary and unpredictable ponds, and the emergence of dormant larvae or pupa of insects inhabiting unpredictable environments, which is distributed over a number of years.

Constant germination and survival of dormant seeds. We assume that a constant fraction G of the seed population in the soil, N, germinate every year, and that a fraction S of the dormant seeds in the soil survives every year. The yield of seeds per unit area for a particular strategy is a random variable K, and the seed yield per plant is proportional to the reciprocal of the density of the plants, i.e. $Y = K/GN$.

If the annual seed yield per plant, Y, as averaged over all the plants, is constant between years, this is the fitness criterion which is maximised by natural selection.

On the other hand, when mean annual seed yield varies between years, the fitness criterion has to take into account the time delay and the averaging caused by the emergence of seedlings from the seed bank in the soil.

Calling $H = (1 - G)S$, we get the recursive expression for the seed population of a single type in the soil:

$$N(t + 1) = N(t)H + K(t) \tag{6}$$

Over a long time, the seed bank population in the soil reaches an equilibrium density N', which is the sum of the surviving dormant seeds in the soil from all the seed yields in the past:

$$N' \simeq \widehat{K}/(1 - H) \tag{7}$$

For a rare type with a strategy yield Y_i, competing with a common type with strategy yield Y, the change in the seed population of strategy i in the soil is:

$$N_i(t + 1) = N_i(t)Y_i/(N_i(t)Y_i + N(t)Y) + N_i(t)H \tag{8}$$

Assuming that N is at its equilibrium level N' from Equation (7), and neglecting N_i in the denominator, we get:

$$N_i(t + 1) = N_i(t)((1 - H)Y_i(t)/Y(t) + H) \tag{9}$$

Long term fitness of the rare type i is therefore:

$$W_i = E(\log((1 - H)Y_i/Y + H)) \tag{10}$$

Expanding W_i by Taylor series around $E(Y_i/Y)$, and neglecting terms with higher derivatives, we get the approximation:

$$W_i = \log((1-H)E(Y_i/Y)+H)-V(Y_{ri})((1-H)^2/(2(H+(1-H)E(Y_i/Y))^2)) < 0 \tag{11}$$

as the condition for noninvading by the rare strategy, where $V(Y_{ri})$ is the variance of Y_i/Y.

In plants with a very effective seed bank, G may be 0.1 to 0.2, and S 0.8 to 0.9. In such cases $(1 - H)^2$ is very small, so that the second term in (Eq. 11) can be ignored. In this case, the criterion for fitness is mainly the mean relative yield $E(Y_i/Y)$, and not the mean logarithm of the ratio, $E(\log(Y_i/Y))$.

Ignoring the second term in (11), the condition for noninvading by a rare strategy becomes approximately:

$$\log((1 - H)E(Y_i/Y) + H) < 0 \tag{12}$$

Inequality (12) allows a stable coexistence between any two strategies or species i and j for which both $E(Y_i/Y_j) > 1$, and $E(Y_j/Y_i) > 1$.

If Y_i and Y_j are independent, then:

$$E(Y_i/Y_j) = E(Y_i)/\text{Har}\ (Y_j) \tag{13}$$

where Har indicates the harmonic mean: $\text{Har}\ (Y) = 1/(E(1/Y))$

By Taylor's expansion:

$$\text{Har}\ (Y) \simeq E(Y)/(1 + V(Y)/E(Y)^2) \tag{14}$$

Thus, the approximate condition for coexistence is that both:

$$(E(Y_i)/E(Y_j))(1 + V(Y_j)/E(Y_j)^2) > 1$$

and

$$(E(Y_j)/E(Y_i))(1 + V(Y_i)/E(Y_i)^2) > 1 \tag{15}$$

The coexistence range given by (15) increases if the difference between the means of Y_i and Y_j decreases and if their variances increase. Coexistence is guaranteed if the mean yields of the two strategies are the same.

If Y_i and Y_j are not independent, $E(Y_i/Y_j)$ decreases if the correlation between Y_i and Y_j is positive, which decreases the coexistence range. A negative correlation has the opposite effect of increasing the coexistence range. No coexistence is possible if Y_i and Y_j are strictly correlated. (See Appendix).

These results demonstrate that when there is a dormant seed bank in the soil, the variance of fitness provides a selective advantage for the rare type when the fitnesses of the two types vary independently to some extent. Fitness independence has the ecological interpretation of differential utilisation of limiting resources or space, which is usually assumed to promote coexistence by the increased competition suffered by the more common types in their favoured niches.

Our results provide a simple model for coexistence in variable environments, which has been analysed more thoroughly by Chesson (1985) and by Shmida and Ellner (1984). The seed bank population in the soil provides the equivalent for overlapping generations or the "Storage Effect" required by the models of Chesson. See also Section II.

Germination is restricted to a small number of favourable years, in between years in which no germination takes place. We assume 100% germination on few rare occasions which occur with probability P, without any germination in the intervening years. In this case, for a single type:

$$N = K \text{ with probability } P \text{ in a germination year} \tag{16}$$

$$N(t) = N(t-1)S \text{ with probability } (1-P) \text{ in other years} \tag{17}$$

The growth of a rare type with Y_i/Y relative seed yield, over an interval of k years between germination years is given by:

$$N_i(t+k) = N_i(t)V^k Y_i/Y \text{ with probability } P(1-P)^k \tag{18}$$

Long term fitness of the rare type is thus:

$$Wi = E(P(1-P)^k \log(V^k) + \log(Y_i/Y)) \tag{19}$$

where the expectation is over all levels of k and Y_i/Y, weighted by their probabilities.

The fitness criterion for strategy i is proportional therefore to $E(\log(Y_i/Y))$, as in the case without dormancy. This is caused by the synchronisation of germination of all the seeds in the same year. It demonstrates that dormancy per se is not the cause of the change of the fitness criterion from the expected log of the ratio to the arithmetic mean of the ratio. Rather, it is caused by the spreading of the germination and yield of the seed yield of any one year over many independent years.

In this case of synchronised germination, the variance in fitness does not allow coexistence, because the expectation of $\log Y$ is a strictly independent measure of long term fitness.

II. Perennial plants

Overlapping generations. We make the following assumptions:
1. There is a constant number of habitable sites per unit area.
2. A fraction of the occupied sites are vacated annualy by the random death of individual plants during the nongrowing season. Annual survival probability of adult plants is S, which may vary randomly between years.
3. A strategy i results in a mean annual seed production per plant per year Y_i, which varies randomly between years.
4. Seeds disperse uniformly over the whole area, and germinate 100% every year.
5. Vacated sites are occupied by different types of seeds according to the Lottery Model (Chesson 1985), with a probability equal to their proportions among all the seeds at the site. Plants reach their adult size during their first year of growth.

This model represents also quite well the life history of most relatively long living iteroparous animals with unsynchronised reproduction.

For any rare strategy i, the annual growth in density in any year t, is:

$$N_i(t+1)/N_i(t) = R_i(t) = S_t + (1-S_t)Y_i(t)/Y(t) \tag{20}$$

N_i decreases in the long run if:

$$E(\log R_i) = E(\log(S + (1 - S)Y_i/Y)) < 0 \qquad (21)$$

Inequality (21) is the condition for noninvading by the rare type Y_i.

Assuming for the time being that S is constant, we expand $E(\log R_i)$ by Taylor's series around $E(Y_i/Y)$, and neglect terms of higher order than the second derivative. By this approximation we get that a rare mutant cannot increase, if:

$$\log(S + (1 - S)E(Y_i/Y)) - \frac{1}{2}V(Y_i/Y)\left(\frac{1-S}{S + (1-S)E(Y_i/Y)}\right)^2 < 0 \qquad (22)$$

where $V(Y_i/Y)$ is the variance of Y_i/Y.

In long living perennial plants, S is close to 1. In this case, the second term in Inequality (22) can almost be neglected. A more exact condition for neglecting the second term is when $V(Y_i/Y)(1 - S)^2$ is much smaller than $2(S + (1 - S)E(Y_i/Y))^2$. A larger S and a smaller ratio between the variance and the mean of Y_i/Y, favour $E(Y_i/Y)$ as the fitness criterion of the rare type.

We conclude therefore that the stability of a strategy against invasion by rare alternative strategies in long living perennial plants is well approximated by the *long term mean relative annual seed yield per plant, and not by the mean logarithm of the annual seed yield.*

This result can be understood intuitively by noting that each individual plant contributes seeds to the reestablished vacant sites from seed yields of many years over its whole life. Thus, it is the averaged relative seed yield over many years which determines the lifetime fitness of the plants.

In this case, a stable coexistence of any two strategies i and j is possible under the same conditions for coexistence in the previously discussed case of a very large seed bank: i.e. if both $E(Y_i/Y_j) > 1$, and $E(Y_j/Y_i) > 1$. Using the same derivations as in the case of the seed bank, we find that if Y_i and Y_j are independent, the approximate conditions for coexistence of different strategies in perennial plants are as given by Inequality (15), i.e. that both:

$$E(Y_i)/E(Y_j)(1 + V(Y_j)/E(Y_j)^2) > 1$$

and

$$E(Y_j)/E(Y_i)(1 + V(Y_i)/E(Y_i)^2) > 1$$

The range of coexistence increases when the variance of both strategies increase, and when the mean yields of the two strategies become more similar.

As in the case of seed dormancy, if Y_i and Y_j are not independent, $E(Y_i/Y_j)$ decreases if the correlation between Y_i and Y_j is positive, which decreases the coexistence range, with the opposite effect of a negative correlation which increases the coexistence range (See Appendix). No coexistence is possible if Y_i and Y_j are strictly correlated.

These results demonstrate that as in the case of the soil seed bank, the variance in fitness provides a selective advantage for the rare type in long living perennial plants if the fitnesses of the two types are independent to some extent. Fitness independence has the ecological interpretation of differential utilisation of limiting resources, or differential responses to other limiting factors, which is usually assumed to promote coexistence by the increased competition suffered by the common types in their favoured niches.

In the case of the perennial plants, the competition between seeds of the same strategy for the occupation of vacant sites is stronger in the more common type if the fitnesses of the different types are at least partly independent, which creates an advantage for the rare type.

If both S and Y are random variables, Taylor's Expansion in two variables gives:

$$
\begin{aligned}
E(\log(S + (1-S)Y_i/Y))) \simeq &\log(\widehat{S} + (1-\widehat{S})E(Y_i/Y)) \\
&- [V(Y_i/Y)(1-\widehat{S})^2 + V(S)(1-E(Y_i/Y))^2 \\
&+ 2\operatorname{Cov}(S, Y_i/Y)]/(2(\widehat{S} + (1-\widehat{S})E(Y_i/Y))^2)
\end{aligned} \tag{23}
$$

Thus, a positive covariance between the relative yield of a rare strategy and adult survival decreases the fitness of the rare strategy. This is because with a higher survival there are fewer vacant sites for the establishment of the seeds.

Rare catastrophic mortality and reestablishment. In this case we assume that all the adult plants are killed simultanously with probability $1 - S$ by some catastrophe such as a fire or a hurricane.

The change in density of the rare type is now:

$$
\begin{aligned}
N_i(t+1)/N_i(t) &= 1 \text{ with probability } S \\
&= Y_i(t)/Y(t) \text{ with probability } 1 - S \tag{24} \\
E(\log R_i) &= (1-S)E(\log(Y_i/Y)) \tag{25}
\end{aligned}
$$

which has the same fitness criterion as the annual plants. We see therefore that the fitness criterion is not determined by the growth habit of the plants, whether annual or perennial. Rather it is determined by the degree of overlap between generations. The mean log criterion is applicable to nonoverlapping generations, which do not allow coexistence, while the mean relative fitness criterion is applicable to overlapping generations, which allow coexistence.

III. Individual stochastic variation in fitness

As shown in Section I, in a well mixed population with wide dispersal, the fitness of any strategy in one generation is measured by the mean fitness of all the individuals with that stategy, and is independent of the individual variation of fitness. It is therefore indifferent to risk, i.e. it is neither risk averse nor risk prone.

The fate of a single mutant or coloniser. There are important biological situations in which the fitness of a new strategy or of a population colonising a new habitat depends on the survival of a small population initiated by a single founder (Gillespie 1977). In such cases, the probability of establishment of a new mutant strategy or of a colonising population depends on the entire probability distribution of fitness for each individual and on the number of such individuals, and not only on their mean fitness, even in constant environments.

For any strategy i, we assume a given fitness probability distribution: $P(k)$, $k = 0, 1, 2, \ldots$ of surviving adult offspring per individual adult, with mean fitness: $\widehat{Y}_i = \Sigma_k P(k)k$.

The ultimate probability of extinction of a single founder and all of its offspring, $P(E)$, is the implicit solution to the equation:

$$P(E) = P(0) + P(1)P(E) + P(2)P(E)^2 + \ldots = \Sigma_k P(k)P(E)^k \qquad (26)$$
$$P(E) = 1 \text{ if } \widehat{Y}_i \leq 1$$

It can be shown that the probability of extinction usually increases if the variance of Y_i increases for the same mean $\widehat{Y}_i$.

Gillespie (1977) has shown by a small variance approximation that the probability of survival is approximately equal to the mean log reproductive success of each individual. This approximation cannot be used in practice, however, because it gives a zero survival probability in all the common cases when there is some probability for zero offspring.

A simple analytical solution for $P(E)$ if the number of offspring is no greater than 2 is the solution to the quadratic equation:

$$P(E)^2 P(2) - (1 - P(1))P(E) + P(0) = 0 \qquad (27)$$

For example, assuming a constant $\widehat{Y}_i = 1 + r$, we get:

$$P(E) = P(0)/(P(0) + r) \qquad (28)$$

$P(E)$ is an increasing function of $P(0)$, which represents the effect of an increasing variance. It is a decreasing function of r, which represents the increase of the mean Y_i, above 1.

The probability of establishment of rare advantagous mutants. Mutants with less variable individual fitness and the same mean fitness will have a higher probability of establishment following an environmental change. This effect becomes important if the establishment probabilities and the numbers of new favourable mutants per generation in the population are very low. Under such conditions, the low variance mutants will be the first to establish and spread in the population following an environmental change. See Gillespie (1977).

The mean fitness is the sole criterion of fitness, however, when the number of the mutants in the population is large enough. In the long run, therefore, the mutants with the highest mean fitness are expected to get established and to displace and succeed all the other mutants.

Each environmental change is expected therefore to be followed by a succession and spread of rare favourable mutants: The first successful mutants are expected to have the highest establishment probability, possibly with a low variance of offspring numbers. Eventually however, high mean and low establishment probability mutants are expected to spread and displace those with a lower mean.

In naturally rapidly changing environments, however, new mutants are probably replacing old alleles all the time. The appearance, establishment and spreading of rare new favourable mutants may be determined therefore by their mutation rates and by their probability of establishment more than by their mean fitness. These conditions will favour mutants with a lower mean and a lower variance, which may have a higher probability of establishment than mutants with a higher mean fitness and a higher variance.

Colonisation of empty habitats by single founders. The same arguments apply to colonisation of empty habitats by one or very few founders. The probability of establishment of a population by a single founder increases if the mean reproductive success increases and the variance decreases. A newly isolated island which is colonised by the very rare arrival of single immigrating individuals, is more likely to be colonised first, therefore, by those species or genotypes which have a higher probability of establishment, which combine a low variance of reproductive success with a high mean. In the long run, however, species with a lower probability of establishment and a higher mean reproductive success will eventually establish, and displace the species with lower mean reproductive success and a lower variance which were established first.

The mean and variance of individual fitness in structured populations in a constant environment. Many natural populations consist of many isolated groups or colonies which are founded by single founders every year at the beginning of the growing season, which grow and disperse by the end of the season (for example aphids and other seasonal insect herbivoures, fungal colonies, bumblebee colonies etc.). In such situations, the probability of establishment of a colony by a single founder is an important component in the fitness of a strategy every year.

To analyse this system, let us make the following assumptions:
1. There are many potential colonisation sites, most of which are unoccupied because the total rate of establishment of new colonies is low.
2. Founding of new groups occurs by individuals in a well mixed population of dispersers resulting from growth and subsequent dispersal from last year colonies.
3. Only one founder arrives at each site while most sites remain empty, because of the relatively low population density of the founders.
4. The amount of resources in a site K is completely utilised during the growing season.
5. Total population density is regulated at some other stage in the life cycle, e.g. by density dependent parasitism.

Calling N the density of a resident type sites, and N_i the density of a rare type, we get:

$$N_i(t+1) = KY_iN_i(t)Cb(1 - P(E)_i) \qquad (29)$$
$$N(t+1) = KYN(t)Cb(1 - P(E)) \qquad (30)$$

where C is a density regulation function such that $N + N_i = 1/C$, b is a survival parameter, $P(E)$ is the extinction probability of the colony at its early stages, and Y

is the mean conversion coefficient of resources for growth and reproduction, which may vary in different alleles.

The relative annual growth rate of the i type is:

$$R_i = (1 - P(E)_i)Y_i/((1 - P(E))Y) \qquad (31)$$

A rare type i cannot invade a common j type if $(1 - P(E)_i)Y_i < (1 - P(E)_j)Y_j$ for all i types, and the winning strategy maximises $(1 - P(E))Y$, without coexistence.

The probability of extinction $P(E)$ is an increasing function of the variance and a decreasing function of the mean of the individual fitness. If the conversion coefficients are identical, the probability of extinction of a single founder is the only factor which determines the long term fitness of a strategy. In this case the winning strategy is expected to maximise the probability of establishment, by an optimal tradeoff between the mean and the variance of the individual fitness.

The model of a structured population can be extended to include the case of persisting colonies which become extinct with probability $1 - S$. In this case also, as long as new colonies are formed by single founders, long term fitness is determined by the probability of establishment of a single founder, which decreases if the variance of individual fitness increases. The winning strategy has a similar optimal tradeoff between the mean and the variance of individual fitness which maximises the probability of establishment.

IV. Random variation of individual fitness components in large populations

Survival and fecundity. The fitness of a strategy is a function of survival and reproduction at different stages of developement or at different environmental conditions in the life of each individual.

The optimal trade off between reproduction and survival is expected to take into account only the mean survival if survival is distributed independently in a large population, e.g. risk of predation or of mortality during dispersal. In such cases, the optimal tradeoff strategy may be for each individual to accept very high risks of mortality by predation or by other factors, if they are balanced by a sufficiently high reproductive success of the surviving individuals.

This is important because it demonstrates an essential difference in the expected risk taking behaviour between animals and humans. A rational human decision maker is not usually expected to choose economic or personal risks which have a significant probability of total ruin or death for *himself*, even if they provide very high returns for the successful gamblers. The human subjective utility function is concerned with satisfying the needs and desires of each individual. It is analogous therefore to the survival probability fitness function which operates on very rare individual new mutants or colonisers, which may be very different from the more common mean fitness criterion in large populations.

This formal distinction between the driving mechanisms of human and animal risk taking behaviour demonstrates that it is impossible to predict human risk taking behaviour by using animal analogies and vice versa. This reduces the usefulness of "sociobiological" theories of human risk taking behaviour. For the same reasons, great care should be taken when applying intuitive or formal models of human behaviour to explain animal risk taking behaviour.

[Note however that such human risk avoiding decision rules do not seem to apply to the decisions of generals and politicians that expose large populations to some very high individual risks.]

The variance of survival reduces long term fitness in large populations only when mortality is synchronised by some environmental catastrophe, or when it occurs in synchronised ages or stages in nonoverlapping generations. In such cases, there is strong selection in favour of strategies with smaller variance of survival, which is expected therefore as part of the natural tradeoff between survival and fecundity.

Similar considerations apply when the variance of survival increases the probability of extinction of rare mutants or founders.

With synchronised survival and fecundity schedules which depend on age or on the stage of developement in nonoverlapping generations, the mean lifetime reproductive output per individual with strategy i in any one generation is:

$$R_i = \Sigma_x L_x B_x \tag{32}$$

where L_x is the mean age or stage dependent survival to age x from birth, and B_x is the mean fecundity at age or stage x.

Since L_x is the product of the mean age dependent survival coefficicients, S_x, from age 1 to age x, it is the geometric mean of the mean age dependent survival raised to the power x. L_x is a decreasing function therefore of the within generation variance of the mean age dependent survival coefficents. Selection favours therefore a lower variance of the mean age dependent mortality factors. Note however that these are population means of the age dependent survivals, which are independent of the individual variance in these parameters.

Similar arguments apply to age dependent mortality caused by randomly occuring environmental stresses which affect the whole population before the age of maturation. In this respect the optimal strategy is expected to have an optimal tradeoff between the mean and the variance which maximises the geometric mean survival until maturation.

On the other hand, the mean reproductive output of individuals adds up arithmetically during their lifetime, weighted by the mean survival. As pointed out already in Section II for long living perennial plants, a rare type can increase if its mean relative fitness is greater than 1. This also gives a selective advantage to a rare type if its age dependent reproductive output is independent to some extent from that of the common type, and allows coexistence.

In either case, in a well mixed population, an optimal strategy maximises mean lifetime reproductive output per individual, and is independent of the individual variance, i.e. lifetime fitness is risk neutral. The individual variation of fitness has an important effect only in cases where the establishment of a new genetic line or a new colony starts by a single or very few founders, as shown in Section III.

Feeding and foraging in varying environments. Feeding sites differ in the mean and variance of the food intake they provide to foragers. A major question in behavioural ecology is whether the fitness of feeding animals is expected to increase or decrease as a function of the individual variance of their food intake (Real and Caraco 1986).

Natural selection is expected to have caused animals to attempt to increase their fitness by prefering the low variance alternatives and be risk averse if the variance *decreases* their fitness, or to prefer the high variance alternative and be risk prone if the variance *increases* their fitness (Real 1980).

There are a number of ways by which the variance in feeding rate at different times affects the long term fitness. Since in large populations, it is only the mean of the individual fitness of a strategy which responds to selection, the effects of the feeding variance have to be assessed only on the mean fitness of the individuals.

According to Real (1980), and Real and Caraco (1986), the mean individual fitness decreases as a function of the feeding variance, and the animals are expected to be risk averse when the fitness function of food intake is convex, i.e. with a decreasing slope. Conversely, the mean individual fitness increases as a function of the variance and the animals are expected to be risk prone when the fitness function is concave, i.e. with an increasing slope.

Some of the observed feeding behaviour of birds at different levels of food deprivation appear to be explained by this model. The birds preferred the low variance alternative when they were relatively well fed, and thus at the convex range of their fitness function, and preferred the high variance alternative when they were partly starved, and presumably at the concave range of their fitness function (Real and Caraco 1986).

Note however that the different risk taking strategies result from the convexity or concavity of the function relating fitness to food intake, and not from the effects of the variance of the fitness component itself.

In addition, the variance in feeding rates decreases the long term fitness more strongly, and more risk averse behaviour is expected, when the variation is correlated among all the individuals in the population, and the generations are less overlapping.

As indicated in the previous section, however, the averaging of the mean survival component of fitness over a number of synchronised occasions within the same generation has to take the geometric mean or the mean log survival. The arithmetic mean, on the other hand, is appropriate for the averaging of the fecundity component of fitness.

Thus, animals could be expected to be more risk averse when survival is the major component of fitness, and more risk prone when fecundity is the major component. Also, a strategy can be less risk averse at a later age, after a large fraction of the reproductive potential has already been materialised (Yoshimura and Clark 1991, Amir and Cohen 1990).

On the other hand, however, the fitness functions are expected to be much more concave when the energy status of the animals is low, and that is the range at which survival becomes important. The two functions have opposite effects, which makes it difficult to make strong predictions, and may also explain the differences reported in the literature.

Conclusions: "What is selected for when fitness varies?" We conclude therefore that the changes in fitness and fitness components take place at very different scales of time, space and population, i.e:

  1. Different events and fitness components during the lifetime of a single individual.

2. Different events and fitness outcomes in the life of different individuals with the same genotype or strategy, relative to the fitness of individuals with other genotypes or strategies.
3. The effect of the temporal, spatial, and demographic structure of the population: i.e. overlapping vs. nonoverlapping generations, clonal vs. nonclonal growth, extinction and recolonisation of patches or colonies, dormancy, dispersal and migration.
4. The effects of the temporal and spatial scales of changes in the environment: i.e. changes within years and generations, between years and generations, and in patches of different sizes and dynamics.

As demonstrated in the models, I have shown that each level or scale of variation requires a different appropriate averaging function to represent long term fitness and selection, which is necessary for calculating or predicting long term evolutionary changes or equilibrium when fitness varies. The overall averaging of fitness has to include therefore all the varying components and their interactions and correlations, each at its appropriate scale.

V. Variance reducing life history adaptations

Natural selection is expected to favour those components of life history strategies which reduce the variance of fitness between generations or years, especially with nonoverlapping generations, and between rare individual founders.

Dormancy and spreading of germination of seeds and spores. The spreading of germination over a large number of years of the seeds produced by annual plants in any one year reduces the annual variance in fitness of these seeds. Seed dormancy establishes a large seed bank population in the soil, which converts nonoverlapping generations of seed production to overlapping seed bank generations. The averaging of fitness over many years reduces the effective variance, increases the long term fitness and is therefore selected for. The selection for seed dormancy increases in more variable environments (Cohen 1966, 1967; Ellner 1985).

This effect of seed dormancy is strong in annual plants. It decreases markedly in perennial or clonal plants with persisting overlapping generations. Seed dormancy is expected therefore to be much more strongly developed in annual plants, and to decrease in importance in perennial or clonal plants with increased longevity.

The perennial life form in plants. The long lived perennial or clonal plant strategy increases long term fitness because it averages the seed yield over many years and in different local patches during the lifetime of each plant or clonal genotype, which reduces the variance of fitness per generation. This selection becomes stronger in environments which are more variable, both in seed yield and in the germination and establishment probability of the seeds (Bulmer 1985).

The reduction of the annual variance and the increased long term fitness in perennial or clonal plants is reduced, however, if there is a long lived seed bank in the soil which buffers the annual variation of seed yield.

Clearly, the averaging effect of the perennial or clonal strategy is not its only contribution to fitness. The averaging effect may be most important in environments where seed yield or establishment are extremely variable between years. For example, many perennial trees or shrubs and clonal plants regenerate by seeds only under exceptional conditions.

Dispersal and migration between patches. Dispersal between patches usually reduces the effective variance of fitness per generation because the seed yields in different patches are independent to some extent. The dispersal and mixing of seeds from different patches reduces therefore the annual variance in seed yield, increases the long term fitness, and is selected for in patchy, spatially independent environments (Levin, Cohen and Hastings 1984).

Dispersal and dormancy of seeds of annual plants are alternative adaptive strategies in randomly varying environments. The optimal level of dispersal decreases when dormancy increases, and the optimal level of dormancy decreases when dispersal increases. The joint equilibrium of dispersal and dormancy in annual plants depends on the relative costs of the two strategies (Cohen and Levin 1987). Plants are expected therefore to specialise either in dormancy or in dispersal of their seeds.

Phenotypic polymorphism. Different phenotypes perform independently to some extent in different environmental conditions. Thus, phenotypic polymorphism of a particular genotype averages to some extent the fitness of the different phenotypes under different conditions. This reduces the variance of fitness of that genotype between generations, and increases its long term fitness in nonoverlapping generations. Typical examples are seed dimorphism in dispersal and germination, and insect polymorphism in flight, diapause, colour, size, behaviour, etc.

Phenotypic polymorphism may also reduce the variance of fitness between patches or events within any one generation, but this will affect the long term fitness only if it increases the mean fitness by a better local or conditional adaptation. Phenotypic polymorphism will promote coexistence if there are complementing density dependent interactions between the different phenotypes under different conditions.

Optimal reduction of clutch or litter size. Parental resources available for raising offspring often vary in different years, and are unpredictable to a large extent by the parents at the time the clutch size has to be determined. The mean number of successfully raised offspring can apparently be maximised by a number of laid eggs which also causes a higher variance in the number of successfully raised offspring. Can this explain the observations that the natural clutch size in some birds is less than needed to maximise the mean number of surviving offspring (Bulmer 1984)?

A clutch size less than needed to maximise the mean number of offspring is expected to increase long term fitness and to be selected for only in short lived birds with strongly correlated fitness in nonoverlapping generations. The individual variance between different breeding pairs within any one year does not influence the mean annual number of offspring, and has no effect on long term fitness. Any strategy which reduces only such individual variation is not selected for in large populations and is not expected in most natural conditions.

The selective advantage of lower annual variance is certainly not expected in long living birds, with considerably overlapping generations, in which the long term fitness is maximised by the mean offspring number per year.

Strategies which reduce the annual variance of successful fledging between years, which are selected for, may also reduce the variance between breeding pairs in the same year by some indirect ecological or behavoural correlation.

Acknowledgement

Supported by US-Israel Binational Science Foundation Grant 89/00130.

References

Amir, S., and D. Cohen. 1990. Optimal reproductive efforts and the timing of reproduction of annual plants in randomly varying environments, *J. Theoret. Biol.* 147:17–42.

Bulmer, M.G. 1984. Risk avoidance and nesting strategies, *J. Theoret. Biol.* 106:529–535.

Bulmer, M.G. 1985. Selection for iteroparity in a variable environment, *Amer. Nat.* 126:63–71.

Chesson, P. 1985. Coexistence of competitors in spatially and temporally varying environments: a look at the combined effects of different sorts of variability, *Theoret. Pop. Biol.* 28:263–287.

Clark, C.W. and J. Yoshimura. 1993. Optimization and ESS analysis for populations in stochastic environments. (This volume).

Cohen, D. 1966. Optimising reproduction in a randomly varying environment, *J. Theoret. Biol.* 12:119–129.

Cohen, D. 1967. Optimising reproduction in a randomly varying environment when a correlation may exist between the conditions at the time a choice has to be made and the subsequent outcome, *J. Theoret. Biol.* 16:1–14.

Cohen, D. and S. Levin. 1987. The intrreaction between dispersal and dormancy strategies in varying and heterogeneous environments. In *Lecture Notes in Biomathematics* (eds. Teramoto, E. & Yamaguchi, M.) 71:110–122.

Ellner, S.P. 1985. ESS germination strategies in randomly varying environments. II. Reciprocal yield models, *Theoret. Pop. Biol.* 28:80–116,

Frank, S.A. and M. Slatkin. 1990. Evolution in a variable environment, *Amer. Nat.* 136:244–260.

Gillespie, J.H. 1977. Natural selection for variance in offspring numbers: a new evolutionary principle, *Amer. Nat.* 111:1010–1014.

Kisdi, E. and G. Meszéna. 1993. Density dependent life history evolution in fluctuating environments. (This volume).

Lacey, E.P., L. Real, J. Antonovics, and D.G. Heckel. 1983. Variance models in the study of life history, *Amer. Nat.* 122:114–131.

Levin, S.A., D. Cohen, and A. Hastings. 1984. Dispersal strategies in patchy environments, *Theoret. Pop. Biol.* 26:165–191.

Lewontin, R.C. and D. Cohen. 1969. On population growth in a randomly varying environment, *Proc. Nat. Acad. Sci. U.S.* 62:1056–60

Maynard Smith. J. 1982. *Evolution and the Theory of Games*. Cambridge University Press, Cambridge.

Orszak, S. 1993. Life history evolution and population dynamics in variable environments: some insights from stochastic demography. (This volume).

Real, L.A. 1980. Fitness, uncertainty, and the role of diversification in the evolution of behaviour, *Amer. Nat.* 115:623–638.

Real, L.A. and T. Caraco. 1986. Risk and foraging in stochastic environments, *Ann. Rev. Ecol. Syst.* 17:371–390.

Real, L.A. and S. Ellner. 1992. Life history evolution in stochasitc environments: A graphical mean-variance approach, *Ecology* 73:1227–1236.

Rubenstein, D.H. 1982. Risk, uncertainty, and evolutionary strategies. In King's College Sociobiology Group (Eds.). *Current Problems in Sociobiology*. Cambridge University Press, pp. 91–111.

Shmida, A. and S.D. Ellner. 1984. Coexistence of plant species with similar niches, *Vegetatio.* 58:29–55.

Slatkin, M. 1974. Hedging one's evolutionary bets, *Nature* 250:704–705.

Stearns, S.C. 1976. Life history tactics: A review of the ideas, *Q. Rev. Biol.* 51:3–47.

Stearns, S.C. 1992. *The evolution of life histories*. Oxford University Press.

Tuljapurkar, S.D. 1989. An uncertain life: demography in random environments, *Theoret. Pop. Biol.* 35:227–294.

Yoshimura, J. and C.W. Clark. 1991. Individual adaptations in stochastic envirnments, *Evol. Ecol.* 5:173–192.

Appendix

When Y_i and Y_j are not independent:

$$E(Y_i/Y_j) = E(Y_i)/\text{ Har }(Y_j) + \text{Cov }(Y_i, 1/Y_j) \qquad (A.1)$$

Noting that for any random variables X and Y:

$$\text{Cov }(X, 1/Y) = E((X - E(X))(1/Y - E(1/Y))) \qquad (A.2)$$

Since by Taylor's expansion:

$$1/Y - E(1/Y) \simeq -(Y - E(Y))/E(Y)^2 + ((Y - E(Y))^2 - V(Y))/E(Y)^3 \qquad (A.3)$$

we get that:

$$\text{Cov }(X, 1/Y) \simeq -\text{Cov }(X, Y)/E(Y)^2 \qquad (A.4)$$

because for symmetrical distributions:

$$E((X - E(X))(Y - E(Y))^2) \simeq 0 \qquad (A.5)$$

Thus:

$$E(Y_i/Y_j) \simeq E(Y_i)/\text{ Har }(Y_j) - \text{Cov }(Y_i, Y_j)/E(Y_j)^2 \qquad (A.6)$$

DENSITY DEPENDENT LIFE HISTORY EVOLUTION

IN FLUCTUATING ENVIRONMENTS

Éva Kisdi

Population Biology Group
Department of Genetics, Eötvös University
Budapest, Múzeum krt. 4/A, Hungary, H-1088

and

Géza Meszéna

Population Biology Group
Department of Atomic Physics, Eötvös University
Budapest, Puskin u. 5-7, Hungary, H-1088

Abstract. Environmental fluctuation may not only alter the life history optimization problem but also query optimization itself. Under density regulation annual growth rate is influenced by the direct effect of fluctuation as well as by an indirect effect due to fluctuating population density. For a weak fluctuation there is an optimal strategy which is slightly different from the stable environment optimum, since (a) it should adapt to an average density altered by fluctuation, (b) it should diminish the fluctuation in annual growth rate caused by direct and indirect effect of environmental fluctuation, and (c) it should exploit an increase, but avoid a decrease in average annual growth rate caused by fluctuation. The "optimal" strategy becomes meaningless if the fluctuation is strong, because long run growth rates are not independent of the established population. Coexistence (or exclusion of the rare strategy) may be mediated by a sufficiently strong fluctuation, which is illustrated by a simple model elucidating the connection with resource-competition models. Moreover, some other consequences of strong fluctuation are demonstrated by the example of a lottery model, such as multiple ESS, ESS which cannot invade an established population, and historical events which determine the outcome of the evolution.

Keywords: coexistence, convergence stability, density dependence, ESS, life history evolution, lottery model, optimality, pairwise invasibility analysis

I. Introduction

Species have to adapt to stochastically varying environments. Despite the substantial amount of variation, life history theory usually considered environmental fluctuation merely as noise around a "true" value. Assuming constant parameters, theoretical models may simply ignore the fluctuation - hence they investigate adaptation in an imaginary stable, "frozen" world. On the other hand, field studies cannot avoid facing the problem of stochasticity. Practical reasons (limited number of data), however, often necessitate the aggregation of data, hence most of the field studies are tempted to consider the average values of the fluctuating parameters to be "true." If we try to learn the stochastic nature from stable environment models and averaged field data, we have to be aware of some difficulties, ranging from numerical discrepancies to new adaptive mechanisms, moreover, to querying adaptation itself.

(1) There may be a simple numerical discrepancy between the "best" strategy in a stable vs. in a fluctuating environment (Schaffer 1974, León 1983). In a population with no age structure and no density dependence the annual growth rate of a strategy, $\lambda(t)$, is influenced by the fluctuating environmental parameter only. Therefore the geometric mean of annual growth rates (or the arithmetic mean of $\ln \lambda(t)$), which governs the long run changes of population composition (cf. Bulmer 1985), can be easily calculated. Since the geometric mean is lowered by a variance in $\lambda(t)$, the "best" strategy should avoid the variability of $\lambda(t)$. If, for example, juvenile survival is affected by fluctuation, it is advantageous to invest more into the constant parental survival and less into the reproduction of uncertain outcome (Schaffer 1974). Mathematically, here lowering the number of offspring decreases the variance of $\lambda(t)$. Heuristically, it introduces a risk spreading: a higher parental survival ensures a longer life span, more chances for reproduction, hence an unsuccessful attempt can be compensated by successful future reproduction events. The same risk spreading effect may motivate species to turn to an iteroparous life history instead of semelparity, when the outcome of reproduction is variable (Murphy 1968, Schaffer and Gadgil 1975).

Risk avoidance, a lower investment into that life history parameter which is affected by fluctuation, is not the only possible means of adaptation. As León (1983) explored, a "promising uncertainty" may induce risk incurrence, an increased effort invested in the fluctuating parameter. If an organism can tolerate bad years without a great loss e.g. in its effective fecundity, but in good years it achieves a high gain, then environmental fluctuation increases the average number of surviving offspring and therefore acts to increase the reproductive investment.

Stochastic models of age structured populations are much more complicated, because annual growth in every year depends on the current age structure, which is also fluctuating in a stochastic environment. Conceptually, however, demography makes no further difficulty, since demographic ergodicity allows assigning a long run growth rate to each strategy independently of the other strategies in the population (Tuljapurkar 1989). The strategy with maximal long run growth rate will spread in the population, just as in the simpler case of no age structure. Risk spreading by iteroparity is known for stochastic demography as well (Goodman 1984, Orzack and Tuljapurkar 1989).

(2) The models considered above implicitly assumed that a population has to cope with environmental stochasticity by applying a single constant strategy. This is not the only possible solution. If the state of the environment is known or can be predicted by the organism when the strategy has to be chosen, the organism can adjust its strategy optimally to the current or foreseeable environment. Adjusting the strategy is inevitably the best solution if the information about the environment is perfect and costless. Using some partial information can also increase the long run growth, while getting more information may be too costly or simply impossible (Cohen 1967, León in this volume). If the relevant information cannot be got when the decision should be made, a further possibility is to adapt a strategy now and make a correction later; e.g. produce a lot of zygotes but eliminate some of them if the actual environment does not allow rearing so many offspring (Temme and Charnov 1987, Kozlowski and Stearns 1989).

Lacking any forecast a genotype (or a species) may adapt to a stochastic environment by mixing several strategies either in a fixed proportion or randomly. Random mixing is known as "adaptive coin flipping," described by Cooper and Kaplan (Cooper and Kaplan 1982, Kaplan and Cooper 1984). Assume that in different years different strategies should be chosen from a discrete strategy set, but the organisms have no guide to choose the appropriate one. Then it may be advantageous if individuals of the same genotype choose different strategies randomly (by "flipping a coin") from the set. Although this mixing cannot ensure such a large annual growth as the pure strategy appropriate in that year could provide, it gives successful individuals of the genotype in each year. Having appropriate and less appropriate individuals in every year, year-to-year variance in annual growth of the mixing genotype is decreased, which is advantageous for the long run growth of the genotype. Similarly, a genotype may generate multiple phenotypes in a precisely fixed proportion; for example, parents may determine the exact frequencies of different progeny types to maximize the long run growth (Yoshimura and Clark 1991).

In case of mixing several phenotypes, the mixing frequencies can be optimized instead of the individual strategy itself. Special problems of this kind are the optimal germination of seeds (Cohen 1966, Bulmer 1984, Ellner 1985a,b) and seed dispersal (Levin et al. 1984, Cohen and Levin 1991). Here the possible strategies are to germinate or to remain dormant; to disperse or not. These discrete strategies should be mixed with an optimal germination or dispersal fraction.

Adaptive coin flipping can be regarded as a discrete version of Levins' fitness set approach originally developed for continuous strategy sets (Levins 1968, León in this volume). According to Levins, strategy mixing may be advantageous if the fitness set is concave - which is the case by definition for a discrete fitness set.

(3) A further, conceptual difficulty comes from density dependence. In density independent models long run growth rates are assigned to strategies (haplotypes or species) independently of each other, hence the "best" strategy having the largest long run growth can be specified independently of the population composition. Under density regulation, however, growth rates of different strategies are not independent parameters. Assume that growth rates of different strategies in a population depend on a single common density measure: adding an individual to the population modifies the growth rate of every strategy independently of the strategy followed by the new individual. The growth rate of each strategy depends on the current common density, which is governed

by all of the strategies in the population. Thus presence and abundance of a strategy may influence the growth of other strategies by altering the common density parameter: long run growth rates are not independent of the other strategies. Although in the simple, frequency-independent stable environment models the concept of the "best" remains valid under density dependence, in a fluctuating environment it becomes simply meaningless to look for the "best" strategy in a strategy set. Instead of the "best," only the "best in a given population" can be defined. Conditionality of long run growths on the population composition has profound consequences like multiple ESS, possible divergence from the neighbourhood of an ESS (lack of convergence stability, Christiansen 1991; cf. Eshel and Motro 1981, Eshel 1983, Taylor 1989, Lessard 1990), coexistence or exclusion of the rare strategy mediated by environmental fluctuation (Levins 1979, Chesson and Warner 1981, Ågren and Fagerström 1984, Bulmer 1985, Ellner 1987, Chesson 1986, 1988, 1989, Chesson and Huntly 1988, 1989).

In the present paper we should like to elucidate some consequences of density regulation on life history evolution in stochastic environments. To concentrate on density dependence, we consider only pure single-phenotype strategies corresponding to haplotypes in a population with no age structure. First we detail the conceptual problems of optimality under density dependence (Section II). In Sections III, IV and V three rather independent models are presented: one can read either of them without any reference to the others. In Section III we investigate how a weak fluctuation alters the ESS compared to the stable environment; this section may be regarded as an extension of León's model (1983) to density dependent populations. In Section IV a simple model is given to illustrate the relation between resource allocation and fluctuation-mediated coexistence of strategies. Finally (Section V), we set up the scheme of a global numerical analysis by which we can search for all the ESS, establish their convergence stability, recognize strategies coexisting or excluding each other and study the possible historical courses of life history evolution in a population with density regulation. The aim of the Discussion is to connect the different approaches presented in the paper, focusing on optimality and coexistence in stochastic environments.

II. Concept of optimality

In a stable environment the annual growth rate λ, which covers the long run growth, is determined by the life history parameters (fecundities and survivals) via the Euler-Lotka equation. Life history parameters in turn depend on the strategy chosen (x), on the actual environment (ξ), and on the actual density (N). Now the "best" strategy maximizing $\lambda(x, \xi, N)$ can be calculated for every N in a given constant environment ξ.

Density is, however, not an independent variable: the equilibrium density $\widehat{N}$ of a homogenous population is determined by the relation $\lambda(x, \xi, \widehat{N}) = 1$. During competition the population is growing while at least one of the competing strategies has an annual growth rate greater than 1. It implies that the strategy with the highest equilibrium density always wins the competition, independently of the initial population composition (Fig. 1; Charlesworth 1980). This "best" strategy has the following particular properties:

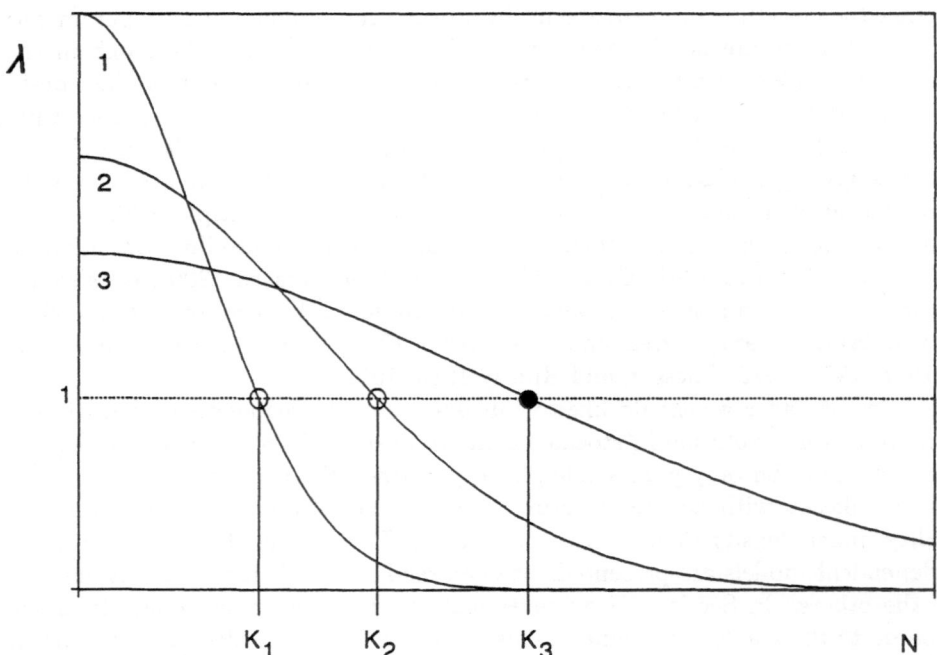

Figure 1. Competition for a single resource in a stable environment. Strategy 3 with the highest equilibrium density (K_3) can initially increase in the equilibrium population of strategy 1 or strategy 2, since its annual growth rate exceeds 1 at density $N = K_1$ as well as at $N = K_2$. Invasion of strategy 3 increases the density, hence the annual growth rate of the former strategy 1 or 2 becomes lower than 1: strategy 3 spreads and the former strategy will be excluded from the population. The population reaches its evolutionary stable equilibrium at $N = K_3$, when strategy 3 is established.

(a) it can initially increase in the established population of any other strategy;

(b) it increases not only initially, but also it spreads until it becomes established excluding the former strategy;

(c) it is an ESS, i.e., its established population cannot be invaded by any other strategy.

We consider the term "optimal" as summarizing features $(a) - (c)$ (cf. the "global ESS" concept of Vincent and Brown 1984; the ESS concept of Ludwig and Levin 1991 comprises (a) and (c)). Feature (a) ensures that there is only one ESS, hence only one optimal strategy. Although the annual growth rate of a strategy is not independent of the other strategies (as it would be in the density-independent case), in a stable environment there exists an optimal strategy because

(i) the interdependence of strategies acts through a single parameter N;

(ii) the annual growth rate of each strategy is a monotonously decreasing function of N; and

(iii) growth of any strategy increases N.

In other words, the concept of optimality is connected to the fact that the population is limited by a single resource. Note that density dependence appears in the models as a simplification: population growth is determined by the abundance of the limiting resource, while resource level is assumed to be a function of density instead of an explicit investigation of resource dynamics. The single density parameter corresponds to a single resource. If more resources limited the population, their level would differently depend on and act on the dynamics of different strategies, hence the growth rates could not be regarded as a function of the overall density. Recall that this is the case in the Lotka-Volterra competition model: there may be k coexisting species on k resources, and the growth rates can be expressed in terms of the k separate densities instead of the resource abundances (Yodzis 1989, p. 120). Lotka-Volterra competition may also make the rare type extinct. Naturally, when several types may coexist or the rare one may be excluded, as it is the case for several resources, the "best type" is not meaningful.

In addition to features $(a) - (c)$, in a continuous strategy set the optimal strategy in a stable environment usually has the following convergence property:

(d) having an established strategy x_e in the neighbourhood of the optimal strategy x_0 and a rare strategy $x_r = x_e + \delta$ with small δ, x_r can invade x_e if and only if x_r is closer to x_0 than x_e; i.e., if $|x_r - x_0| < |x_e - x_0|$.

This latter property means that the optimal strategy is convergence stable (Eshel and Motro 1981, Eshel 1983, Taylor 1989, Lessard 1990, Christiansen 1991); roughly speaking, strategies closer to the optimal one are "better." Convergence stability is ensured by two-times differentiability of the equilibrium density as a function of strategy x.

Environmental fluctuation fundamentally changes the situation described above. In a fluctuating environment long run growth is covered by $\overline{\ln \lambda(t)}$, while the annual growth rate $\lambda(t)$ depends on the current environment ξ_t and on the current density N_t. A fluctuation in the environment, as a direct effect, leads to a fluctuation in one or more life history parameters (Fig. 2). Fluctuation of some life history parameters makes the population density fluctuate; and density fluctuation, in turn, cause a fluctuation in density dependent parameters, as an indirect effect of the stochastic environment. The distribution of population density, hence the indirect effect, depends on the life history parameters of the established population.

In a population with fluctuating density long run growth of a strategy generally depends not only on a single, but on several attributes of density distribution (e.g. mean, variance, and further moments). Different strategies may modify the attributes of the density distribution in different ways: e.g. spread of one strategy may increase, while spread of another may decrease the variance in density. On the other hand, various strategies may respond differently to changes in a density distribution attribute: e.g. increased variance may be advantageous for a particular strategy, disadvantageous for another one. All of the conditions (i)-(iii) are violated, since growth rates depend on more than one parameter, not in a uniform way, and different strategies have different effects on the relevant parameters of density distribution.

Dependence of growth rates on several parameters instead of a single one is analogous to the case of several limiting resources. Indeed, different moments of density distribution can be regarded as different resources (Levins 1979). In Section IV we will present an example where average density and density variance can be paralleled with two resources. As with several resources, in a fluctuating environment there is not an equivalent of the density maximization principle, which could ensure the existence of

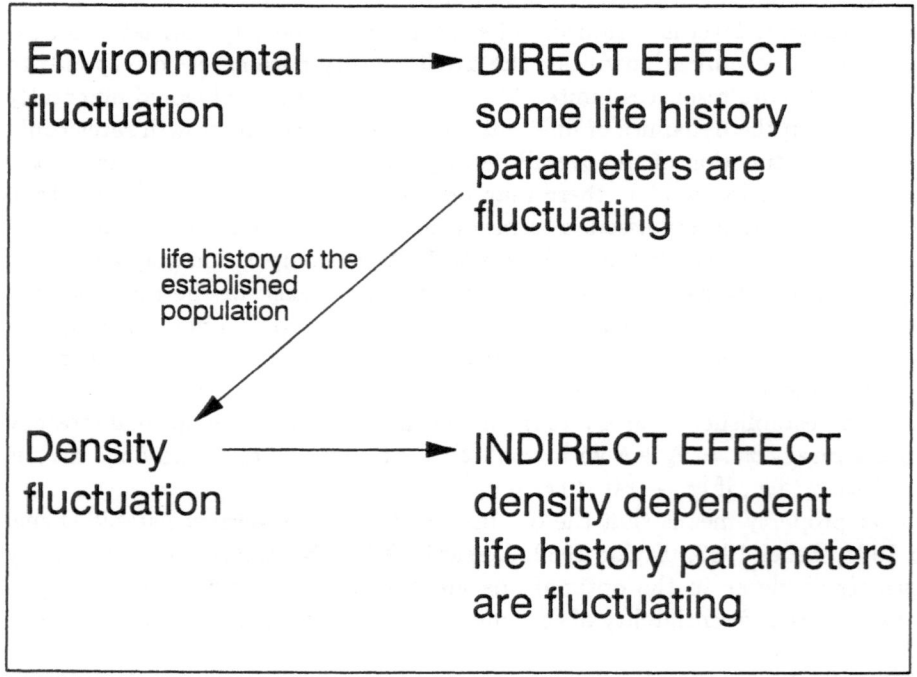

Figure 2. Direct and indirect effects of environmental fluctuation.

an optimal strategy having all of the features $(a) - (c)$. A strategy which can initially increase in a given population may not exclude the strategy established before; instead, they may coexist in fluctuating environment as for the case of several resources. A strategy which cannot be invaded, hence satisfies the ESS feature (c), may be unable to invade an established population of another strategy, which disagrees with (a). Since features $(a) - (c)$ are detached, the concept of optimality generally becomes meaningless.

Lacking an optimal strategy it is straightforward to investigate features $(a) - (d)$ separately. Features (a), (c) and (d) concern only the initial increase of a rare strategy in a given established population, while (b) regards the much more complicated problem of dynamics of mixed populations.

$Ad(a)$. Initial increase of a rare strategy x_r emerging in very low frequency in an established stationary population is covered by its boundary growth rate $\overline{\ln \lambda(x_r, \xi_t, N_t)}$, where averaging is taken on the stationary joint distribution (ξ_t, N_t) of the established population. Investigation of boundary growth rates, called standard invasibility analysis (Turelli 1978), allows us to decide whether a rare strategy can invade an established one or not.

$Ad(b)$. Dynamics of different strategies in a mixed population is a very difficult, unfeasible task for analysis. The most important results have come from Chesson and Ellner (1989), under the restrictive assumption of undercompensating density regulation. They could characterize the evolution in a mixed population of two strategies by their boundary growth rates. Let $\Delta_1 = \overline{\ln \lambda(x_1, \xi_t, N_t)}$ and $\Delta_2 = \overline{\ln \lambda(x_2, \xi_t, N_t)}$ the boundary growth rates of strategy 1 and 2, respectively, in the established population of the other one. Then

- if both Δ_1 and Δ_2 are positive, the population reaches a stationary limit distribution containing both of the strategies, i.e., the two strategies will coexist;

- if $\Delta_1 > 0$ and $\Delta_2 < 0$, strategy 1 outcompetes strategy 2 and finally becomes established excluding strategy 2;

- if both Δ_1 and Δ_2 are negative, starting with a mixed population one of the competing strategies will go extinct.

The first statement was derived by Ellner (1989) without assuming undercompensating density regulation, but making an irreducibility assumption on the dynamics. Chesson (1982) proved all of the three assertions for the lottery model. Although derivation of these statements requires more or less restrictive assumptions, they are usually assumed to be fulfilled in general. Particularly, if two strategies can invade each other's established population, they are considered as coexisting (Turelli 1978, Chesson and Warner 1981, Ågren and Fagerström 1984, Ellner 1987, Chesson 1988, 1989).

$Ad(c)$. The concept of the ESS remains quite meaningful in fluctuating environments as well. By invasibility analysis we can explore strategies which cannot be invaded by any other strategy. Opposed to the case of a stable environment, in a fluctuating environment there may be more than one ESS. It is not sure at all, however, that an ESS can invade an established population of another strategy. Indeed, examples can be given when the ESS emerging in low frequency is excluded from a non-ESS population (Kisdi and Meszéna *in prep.*).

$Ad(d)$. Convergence stability is not necessarily satisfied for an ESS in fluctuating environment. In a specific model we could find three types of ESS: convergence stable which could invade directly its neighbours; convergence stable which directly couldn't invade; and convergence unstable which couldn't invade (Kisdi and Meszéna *in prep.*).

III. How does a weak fluctuation alter the ESS under density regulation?

Here we address the question how a weak fluctuation shifts the maximum of $\overline{\ln \lambda}$ compared to the stable environment; i.e., how the ESS in a weakly fluctuating environment differs from the stable environment optimal strategy. The basic view postulated in this section is that a weak fluctuation leads to the establishment of a single ESS, which is only slightly different from the stable environment optimum.

To build a simple model, consider a population without age structure. Annual growth rate in year t is

$$\lambda(t) = \lambda(x, \xi_t, C_t) \tag{1}$$

where x stands for the strategy chosen from a continuous strategy set (like reproductive effort, number of offspring, some measure of parental care etc.). Environment is characterized by a single variable ξ_t, which is assumed to constitute an ergodic stochastic process. C_t measures the strength of competition. C_t may refer simply to the population density in year t ($C_t = N_t$), but we can allow for more general forms of competition, such as dependence on the number in some critical age group (juveniles or adults; Charlesworth 1980) or on another significant measure like the number of competing juveniles per free space in the lottery model (Box 1). $\lambda(t)$ is supposed to be an increasing function of ξ_t , while a decreasing function of C_t ($\frac{\partial \lambda}{\partial \xi} > 0$ and $\frac{\partial \lambda}{\partial C} < 0$).

To get the long run growth rate of a strategy, we have to average its log annual growth rate on the joint limit distribution (ξ_t, C_t). If an environmental process ξ_t is given, the competition process C_t is determined by the established population. Therefore the boundary long run growth rate of a rare strategy x_r can be calculated for a given environment and a given established strategy x_e. Given an environmental process, denote the boundary growth rate by

$$\Delta(x_r, x_e) = \overline{\ln \lambda(x_r, \xi_t, C_t)} \qquad (2)$$

where (ξ_t, C_t) corresponds to x_e.

By definition, an established strategy x_e is an ESS if no other strategy can invade it, i.e., no other strategy has a greater long run growth rate than the established strategy itself. Consider the long run growth rates of possible rare strategies given that x_e is established (Fig. 3). x_e is an ESS if $x_r = x_e$ corresponds to the global maximum of $\Delta(x_r, x_e)$ for a given x_e (cf. Maynard Smith 1982, p. 26). Hence the ESS has to satisfy the local maximum conditions

$$\frac{\partial \Delta(x_r, x_e)}{\partial x_r} \big|_{x_r = x_e} = 0 \qquad (3a)$$

$$\frac{\partial^2 \Delta(x_r, x_e)}{\partial x_r^2} < 0 \qquad (3b)$$

For comparison, we have to define a stable environment. Here the environmental variable ξ has a constant value, which we let equal to the average $\overline{\xi_t}$ in the fluctuating environment: it means that turning from the stable to the fluctuating environment we introduce a variation into ξ keeping the average unchanged. Let x_o be the optimal strategy in the stable environment. x_o determines a constant competition parameter C_o in the stable environment and satisfies the usual optimality conditions

$$\frac{\partial \lambda}{\partial x} \big|_{x_o, C_o} = 0 \qquad (4a)$$

$$\frac{\partial^2 \lambda}{\partial x^2} < 0. \qquad (4b)$$

It is plausible to suppose that a sufficiently weak fluctuation cannot alter the qualitative features of the fitness function; in particular, it cannot make a concave function convex. Hence we suppose that the fluctuation shifts the stable environment optimum into a nearby maximum, which refers to the fluctuating environment ESS; thus we seek for the ESS by looking for the solution of Eq. (3a) in the neighbourhood of x_o and don't check directly that this extremum refers to a maximum, not a minimum. x_o satisfies Eq. (4b) and it should imply Eq. (3b) as well. Similarly, if λ has more than one local maximum, we can suppose that the global maximum in the stable environment, which corresponds to the optimal strategy, shifts into the global maximum in the fluctuating environment, which is the ESS: a sufficiently small fluctuation cannot change the shape of the fitness function enough to alter which maximum is the highest one.

Looking for the ESS in a lottery model

Assume that the population is limited by the number of spaces where the settled individuals can live. An adult keeps its space for the whole life span; juveniles compete for the spaces that become free by adult death. Juveniles are always in surplus and those who failed to find free space die. The previous assumptions cover the lottery density regulation (Chesson and Warner 1981). Here we consider an environmental fluctuation which acts on adult survival as a random multiplier.

Annual growth rate of a rare strategy is given by

$$\lambda_r = \frac{n_r}{C_t} + p(n_r)\xi_t \qquad (B1)$$

where

n_r, n_e =fecundities characterizing the rare and the

established strategies (i.e., $x_r = n_r, x_e = n_e$)

$p(n)$ =average parental survival, traded off with n

ξ_t =random multiplier, $\overline{\xi_t} = 1$

$C_t = \dfrac{n_e}{1 - p(n_e)\xi_t}$.

The strength of competition, C_t, is measured by the number of competing juveniles per free space, or by the inverse of settling probability in an established population of strategy n_e.

In a stable environment the annual growth rate of an established strategy has to be 1, i.e.

$$\frac{n_e}{\widehat{C}} + p(n_e) = 1 \qquad (B2)$$

Calculating the derivatives in Eq. (5) for the annual growth rate $(B1)$ we get

$$\frac{\partial \Delta}{\partial n_r}\Big|_{n_e} = \overline{\left(\frac{1}{\lambda}\right)} \frac{\partial \lambda}{\partial n_r}\Big|_{n_e, \bar{\xi}, \widehat{C}} + \left[-\frac{1}{\widehat{C}^2}\right](\overline{C} - \widehat{C}) + \left[-p(n_e)\frac{\partial p}{\partial n}\Big|_{n_e}\right] V(\xi)$$

$$+ \left[-\frac{n_e}{\widehat{C}^4} + \frac{1}{\widehat{C}^3}\right] V(C) + \left[\frac{n_e}{\widehat{C}^2}\frac{\partial p}{\partial n}\Big|_{n_e} + \frac{p(n_e)}{\widehat{C}^2}\right] \text{COV}(\xi, C) \quad (B3)$$

From the definition of C_t and using $(B2)$ we can approximate $(\overline{C} - \widehat{C})$, $V(C)$ and $\mathrm{COV}(\xi, C)$ as

$$\overline{C} - \widehat{C} = \frac{p^2(n_e)\widehat{C}^3}{n_e^2}V(\xi)$$

$$V(C) = \frac{p^2(n_e)\widehat{C}^4}{n_e^2}V(\xi) \qquad\qquad (B4)$$

$$\mathrm{COV}(\xi, C) = \frac{p(n_e)\widehat{C}^2}{n_e}V(\xi)$$

Finally substituting $(B4)$ into $(B3)$ results

$$\frac{\partial \Delta}{\partial n_r}\Big|_{n_r = n_e} = \overline{\left(\frac{1}{\lambda}\right)}\frac{\partial \lambda}{\partial n_r}\Big|_{n_e, \overline{\xi}, \widehat{C}}$$

All the effects shifting the ESS have been canceled out. The strategy which is optimal in the stable environment is an ESS in a weakly fluctuating environment as well. This derivation has been presented to illustrate the use of Eq. (5); by a direct evaluation it can be shown that in this lottery model the stable environment optimal strategy is an ESS not only for a weak but also for arbitrary strong fluctuation (Kisdi and Meszéna *in prep.*).

To see how the fluctuating environment ESS differs from the stable environment optimal strategy consider the ESS condition in Eq. (3a). By Taylor-expansion in the variables ξ_t and C_t about their values in the stable environment, we can express $\frac{\partial \Delta}{\partial x_r}\Big|_{x_r = x_e}$ in terms of the means, variances-covariance etc. of the (ξ_t, C_t) distribution. Let

$\overline{\xi}$ = constant value of ξ in stable environment

= mean value of ξ_t in fluctuating environment

$\widehat{C}$ = constant value of C if x_e is established in the stable environment

$\overline{C}$ = mean value of C_t when x_e is established in the fluctuating environment

$V(\xi)$ = variance of ξ_t in fluctuating environment

$V(C)$ = variance of C_t in fluctuating environment, given that x_e is established

$\mathrm{COV}(\xi, C)$ = covariance of ξ_t and C_t, given that x_e is established.

If fluctuation is weak in the sense that $V(\xi)$, $V(C)$, $\mathrm{COV}(\xi, C)$ and $(\overline{C} - \widehat{C})$ are small, and the third and higher order central moments of the (ξ_t, C_t) distribution are

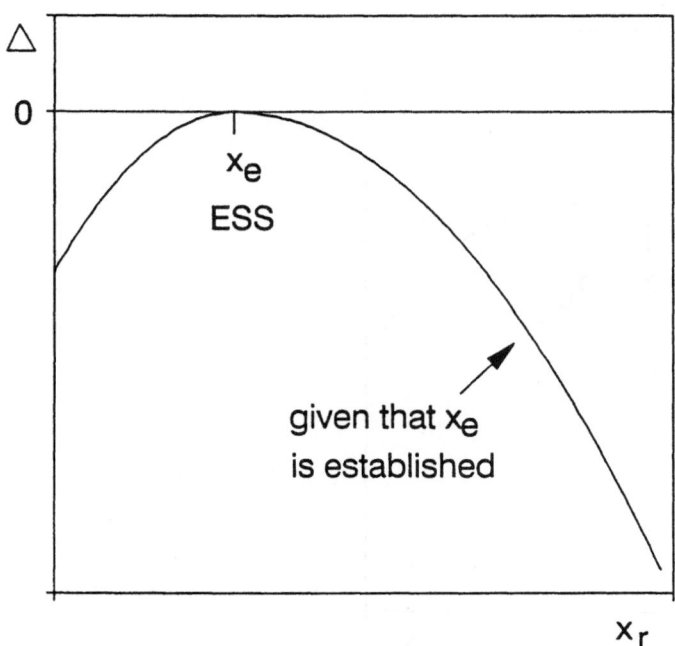

Figure 3. Boundary growth rate $\Delta(x_r, x_e)$ as a function of the rare strategy x_r, given that x_e is established. x_e is an ESS if no other strategy can invade it, i.e., no strategy has a positive boundary growth rate.

negligible, Taylor-approximation gives the following result:

$$\frac{\partial \Delta}{\partial x_r}\big|_{x_r=x_e} = \overline{\left(\frac{1}{\lambda}\right)} \cdot \frac{\partial \lambda}{\partial x}\big|_{x_e, \bar{\xi}, \widehat{C}} + \frac{\partial^2 \lambda}{\partial x \partial C} \cdot (\overline{C} - \widehat{C}) +$$

$$+ \left[-\frac{\partial^2 \lambda}{\partial x \partial \xi} \cdot \frac{\partial \lambda}{\partial \xi} + \frac{1}{2} \cdot \frac{\partial^3 \lambda}{\partial x \partial \xi^2} \right] V(\xi) +$$

$$+ \left[-\frac{\partial^2 \lambda}{\partial x \partial C} \cdot \frac{\partial \lambda}{\partial C} + \frac{1}{2} \cdot \frac{\partial^3 \lambda}{\partial x \partial C^2} \right] V(C) +$$

$$+ \left[-\left(\frac{\partial^2 \lambda}{\partial x \partial \xi} \cdot \frac{\partial \lambda}{\partial C} + \frac{\partial^2 \lambda}{\partial x \partial C} \cdot \frac{\partial \lambda}{\partial \xi} \right) + \frac{\partial^3 \lambda}{\partial x \partial \xi \partial C} \right] \text{COV}(\xi, C) \qquad (5)$$

which has to be zero for the ESS. The derivation of Eq. (5) is given in the Appendix.

First let us investigate the zero order term $\overline{\left(\frac{1}{\lambda}\right)} \cdot \frac{\partial \lambda}{\partial x}\big|_{x_e, \bar{\xi}, \widehat{C}}$. As you may notice, $\overline{\left(\frac{1}{\lambda}\right)}$ depends on the variation in λ; throughout the whole reasoning, however, we need only to know that it is a positive factor. $\frac{\partial \lambda}{\partial x}\big|_{x_e, \bar{\xi}, \widehat{C}}$ refers to strategy x_e established in the stable environment. If $\frac{\partial \lambda}{\partial x}\big|_{x_e, \bar{\xi}, \widehat{C}}$ is positive, then strategies with slightly greater x can invade the established strategy x_e in the stable environment, while strategies with

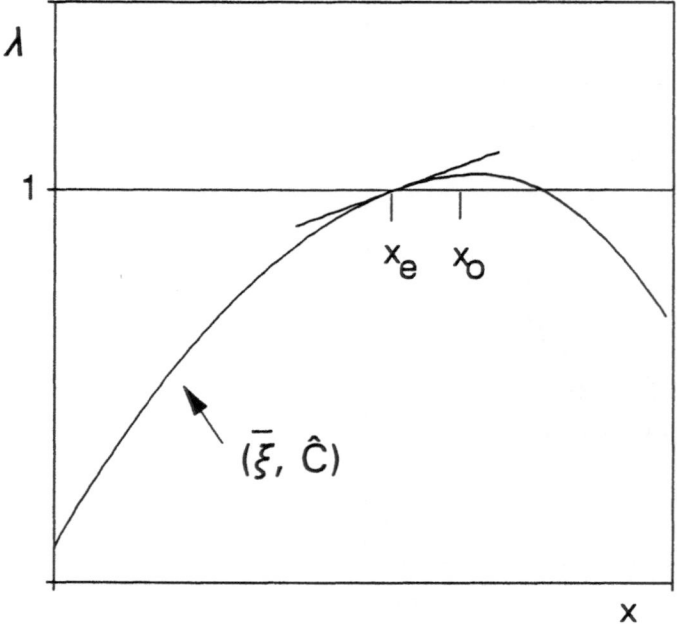

Figure 4. Stable environment annual growth rate as a function of x at the equilibrium density of strategy x_e when it is established in the stable environment. Since x_e cannot invade or exclude itself, $\lambda(x_e)$ has to be 1. If the slope $\frac{\partial \lambda}{\partial x}\big|_{x_e, \bar{\xi}, \widehat{C}}$ is positive, the strategies which can invade x_e in the stable environment are slightly greater than x_e. The stable environment optimal strategy (x_o) must be able to invade, hence it must be somewhat greater than x_e.

lower x cannot invade (Fig. 4).Since x_e is assumed to be in the neighbourhood of the stable environment optimal strategy x_o and x_o can invade all other strategies in stable environment, it implies that x_o has to be greater than x_e. In the same way, a negative $\frac{\partial \lambda}{\partial x}\big|_{x_e, \bar{\xi}, \widehat{C}}$ means that x_o is lower than x_e.

For the ESS, the Taylor-series in Eq. (5) has to sum up to zero. If the sum of higher order terms results in a positive value, the zero order term must be negative, which implies $x_e > x_o$; similarly, higher order terms giving a negative sum lead to $x_e < x_o$. Higher order terms describe the various mechanisms shifting the ESS compared to the stable environment. For investigation of these mechanisms we have just got a guide: every positive term corresponds to a mechanism by which the fluctuation increases the ESS, while negative terms describe mechanisms lowering the ESS compared to the stable environment optimum.

Now we can interpret the higher order terms in turn.

(1) $\frac{\partial^2 \lambda}{\partial x \partial C} \cdot (\overline{C} - \widehat{C})$: *"Use the opportunity of reduced average competition!"*

In a fluctuating environment, an established strategy experiences an average competition $\overline{C}$ usually different from the stable environment competition parameter, $\widehat{C}$, of the same strategy. Moreover, it can be shown that average competition $\overline{C}$ is usually lower than $\widehat{C}$, giving a negative difference in $(\overline{C} - \widehat{C})$, except if λ is a strongly convex function of ξ and C. Environmental fluctuation therefore typically acts to decrease the average strength of competition.

A negative second partial derivative $\frac{\partial^2 \lambda}{\partial x \partial C}$ means that increasing x makes the competition dependence of λ stronger (negative slope $\frac{\partial \lambda}{\partial C}$ becomes steeper). In this case a reason why not to choose a high x is to avoid the strong competition effect. If the fluctuation lowers the average competition, then x may be increased, since the lower competition parameter allows for a stronger dependence. Formally, a negative $\frac{\partial^2 \lambda}{\partial x \partial C}$ with a negative $(\overline{C} - \widehat{C})$ gives a positive term in Eq. (5), which corresponds to a positive shift in x characterizing the ESS.

Generally, a similar effect would be caused by the difference between average and constant environmental parameters in fluctuating and stable environments, respectively. However, this effect has been canceled out, since the two environments were defined to have the same mean value of ξ. A change in average environment or in average competition has the same consequences as in stable environment optimization models (for a joint investigation of environmental and density effects see Pásztor 1988, Meszéna and Pásztor 1990, Meszéna *in prep.*).

(2) Effects of environmental and competition variances

Variances of environmental and competition parameters ($V(\xi)$ and $V(C)$) act in a similar way. Their effects can be separated into two parts. First, they introduce a variance into $\lambda(t)$; and any fluctuation in $\lambda(t)$ around a given mean lowers the long run growth rate, therefore should be avoided (2a and 2c below, respectively). Second, a fluctuation in ξ_t or C_t may change the mean λ, therefore modify $\overline{\ln \lambda}$ as well (2b and 2d).

(2a) $-\frac{\partial^2 \lambda}{\partial x \partial \xi} \cdot \frac{\partial \lambda}{\partial \xi} V(\xi)$: *"Avoid a variance in $\lambda(t)$ caused by fluctuation in the environment!"*

Given a mean λ, any variance in $\lambda(t)$ lowers the long run growth rate $\overline{\ln \lambda}$: the logarithm function being concave, the average log λ is less than the log mean λ (Jensen's inequality, Fig. 5).Any fluctuation in ξ_t leads to a variance in $\lambda(t)$, but the amount of variance in $\lambda(t)$ may be influenced by the strategy chosen. Therefore a strategy avoiding the variability of $\lambda(t)$ would be advantageous.

In the term $-\frac{\partial^2 \lambda}{\partial x \partial \xi} \cdot \frac{\partial \lambda}{\partial \xi} V(\xi)$ the variance $V(\xi)$ measures the strength of fluctuation in ξ_t, while the positive derivative $\frac{\partial \lambda}{\partial \xi}$ refers to the sensitivity of λ to a change in ξ. Variance in $\lambda(t)$ caused by a given $V(\xi)$ can be altered by changing this sensitivity: a strategy which is less sensitive to the fluctuating environmental parameter ξ_t experiences a lower variability in $\lambda(t)$. The second partial derivative $\frac{\partial^2 \lambda}{\partial x \partial \xi}$ measures the effect of x on the sensitivity $\frac{\partial \lambda}{\partial \xi}$. If it is negative, then increasing x decreases the sensitivity $\frac{\partial \lambda}{\partial \xi}$, hence decreases the variance of $\lambda(t)$. With negative $\frac{\partial^2 \lambda}{\partial x \partial \xi}$ the whole term becomes positive, i.e., it acts to increase the ESS: the ESS is shifted upward if the variance of $\lambda(t)$ can be lowered in this way. If increasing x leads to an increased sensitivity ($\frac{\partial^2 \lambda}{\partial x \partial \xi}$ is positive), then x should be decreased to avoid the variability of $\lambda(t)$.

(2b) $\frac{1}{2} \cdot \frac{\partial^3 \lambda}{\partial x \partial \xi^2} V(\xi)$: *"By exploiting good years, achieve a high average annual growth rate!"*

A fluctuation in ξ_t, beside the consequent variance in $\lambda(t)$, may change the mean value of $\lambda(t)$ as well. Take a strategy whose annual growth rate decreases only by a

Figure 5. Jensen's inequality. Take any function $f(u)$ the independent variable of which, u, is fluctuating. The average of $f(u)$ is (a) greater than, (b) equal to, or (c) less than $f(\overline{u})$, if the function $f(u)$ is convex, linear or concave downwards, respectively. For (2a) in Section III take $u = \lambda$ and $f(u) = \ln(\lambda)$; the logarithm function being concave, the average $\ln \lambda$ is less than $\ln(\overline{\lambda})$ according to figure (c). For (2b) in Section III replace u by ξ and $f(u)$ by $\lambda(\xi)$; the average λ is (a) increased, (b) isn't changed and (c) decreased compared to $\lambda(\overline{\xi})$ when $\lambda(\xi)$ is (a) convex, (b) linear and (c) concave, respectively.

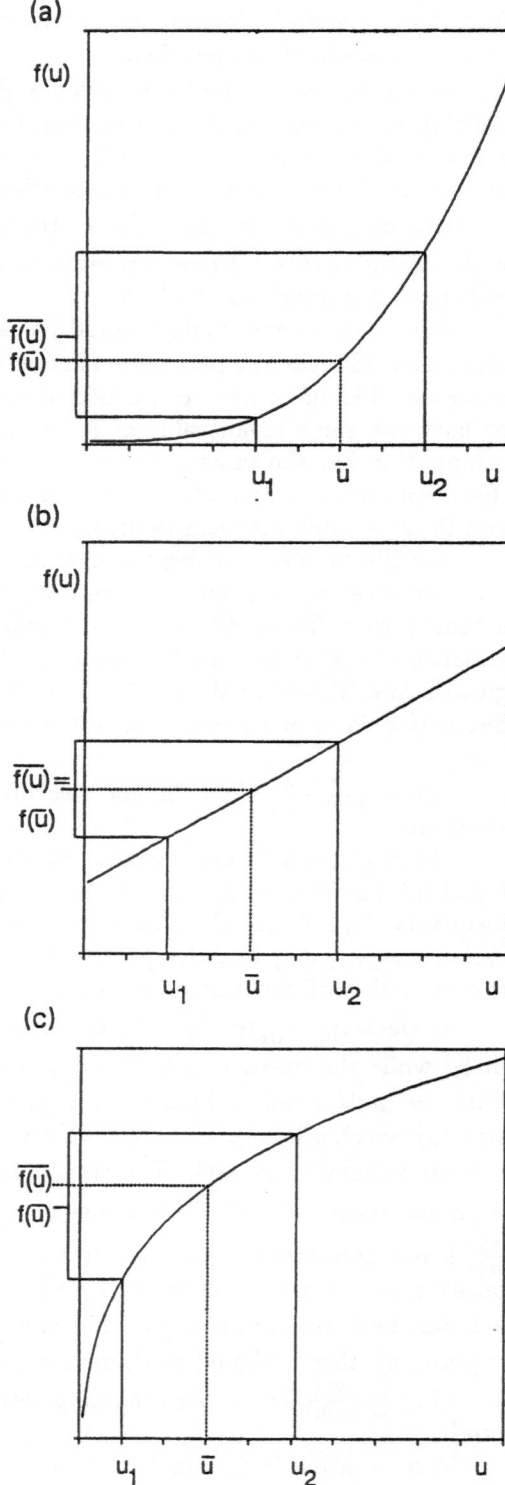

small amount in bad years (when ξ_t is low compared to the average environment), while in good years (when ξ_t is high) it has a high increase in λ. By exploiting good environmental conditions and not suffering from bad ones this strategy achieves a high average annual growth rate. Mathematically, average λ will be increased by the fluctuation of ξ if λ is a convex function of ξ (by 'convex', we mean convex downwards, i.e., $\frac{\partial^2 \lambda}{\partial \xi^2} > 0$ throughout the paper), it will be unchanged if λ is linear in ξ ($\frac{\partial^2 \lambda}{\partial \xi^2} = 0$), while it will be decreased if λ is a concave function of ξ ($\frac{\partial^2 \lambda}{\partial \xi^2} < 0$; Jensen's inequality, Fig. 5). An increase in $\overline{\lambda}$ means an increase in the long run growth rate $\overline{\ln \lambda}$ as well. Therefore it is advantageous to choose a strategy which is associated with a convex function $\lambda(\xi)$.

If convexity of $\lambda(\xi)$ can be increased by choosing a higher x, then a strategy with higher x exploits more the possibility to increase $\overline{\lambda}$: hence a higher x should be chosen. Indeed, if a higher x increases the convexity (or decreases the concavity) of $\lambda(\xi)$, then $\frac{\partial^3 \lambda}{\partial x \partial \xi^2}$ is positive, which acts to increase the ESS compared to the stable environment optimum. If a higher x decreases the convexity (or increases the concavity) then x should be decreased to achieve a higher $\overline{\lambda}$; accordingly, $\frac{\partial^3 \lambda}{\partial x \partial \xi^2}$ is negative.

(2c) $-\frac{\partial^2 \lambda}{\partial x \partial C} \cdot \frac{\partial \lambda}{\partial C} \mathrm{V}(C)$: *"Avoid a variance in λ caused by fluctuation in the strength of competition!"*

Any environmental fluctuation leads to a fluctuation in population numbers, hence in the strength of competition. Fluctuation of the competition parameter C_t modifies the long run growth rate by the same mechanisms as the fluctuating environmental parameter itself. First, a fluctuation in C_t leads to a variation in $\lambda(t)$, which should be diminished by choosing a strategy whose annual growth rate is less sensitive to changes in C (cf. (2a)). If a higher x makes λ less sensitive to C, i.e., it makes the negative slope $\frac{\partial \lambda}{\partial C}$ be less steep, then $\frac{\partial^2 \lambda}{\partial x \partial C}$ is positive. Since $\frac{\partial \lambda}{\partial C}$ is negative by definition, the whole term becomes positive: if a higher x lowers the variance in $\lambda(t)$, x should be increased.

(2d) $\frac{1}{2} \cdot \frac{\partial^3 \lambda}{\partial x \partial C^2} \mathrm{V}(C)$: *"By exploiting low competition years, achieve a high average annual growth rate!"*

A fluctuation in C_t may modify the average annual growth rate $\overline{\lambda}$ in the same way as the environmental fluctuation in (2b). If a strategy does not suffer much from the increased competition, but it can realize a high annual increase in years with low competition, it may reach a high average annual growth rate when competition is fluctuating. Increasing the convexity of $\lambda(C)$ is therefore advantageous. This effect is described by the third derivative $\frac{\partial^3 \lambda}{\partial x \partial C^2}$: if increasing x increases the convexity $\frac{\partial^2 \lambda}{\partial C^2}$ then the third derivative is positive, indicating a positive shift in the ESS.

(3) Covariance effects

Variance terms in (2) alone describe the effect of fluctuation only if the fluctuating parameters ξ_t and C_t are statistically independent. ξ_t and C_t may, however, be correlated for several reasons. This is the case if the ξ_t process, and consequently the C_t process have some autocorrelation. A long run of good years for example allows the population to reach a high density, therefore implies a high competition after a transient increase. Similarly, a long run of bad years leads to low competition. If the environmental process tends to consist of good and bad runs, it results in a positive correlation between ξ_t and C_t. Another important possibility is invoked by Chesson (Chesson and Warner 1981, Chesson and Huntly 1988, 1989). Assume that the fluctuating environmental factor acts early in the season, and modifies the population number in the critical age group before competition. Then the strength of the subsequent competition depends on the environmental factor in the same season. If a good year allows for higher density, there will be a higher competition in good years, hence a positive correlation emerges between ξ_t and C_t. The lottery model (cf. Box 1.) is an example for this mechanism (Chesson and Huntly 1988, 1989). Here juveniles have to compete for spaces which have become free by adult death; if adult survival is affected by some environmental fluctuation, this induces a correlated fluctuation in the number of free spaces, hence in the strength of competition.

Similar to variances $V(\xi)$ and $V(C)$, covariance $COV(\xi, C)$ has a twofold effect on the long run growth rate, modifying the variance in $\lambda(t)$ and shifting $\overline{\lambda}$.

(3a) $-\left(\frac{\partial^2 \lambda}{\partial x \partial \xi} \cdot \frac{\partial \lambda}{\partial C} + \frac{\partial^2 \lambda}{\partial x \partial C} \cdot \frac{\partial \lambda}{\partial \xi}\right) COV(\xi, C)$: *"Use a smoothing, but avoid a variance magnifying effect of covariation!"*

A negative covariance between ξ_t and C_t would mean that there is a tendency to have very good years with good environment and low competition, and very bad years with bad environment and high competition. In very good and very bad years $\lambda(t)$ is very different from its mean value, therefore the variance in $\lambda(t)$ is magnified compared to the independent effects of environment and competition. Conversely, a positive covariance introduces a smoothing effect over years: if good environment - high competition and bad environment - low competition years are frequent, $\lambda(t)$ values are more equalized. Therefore a positive covariance is advantageous, while a negative covariance is disadvantageous regarding the variability of $\lambda(t)$.

The impact of covariance is increased, if λ is more sensitive either to the environmental or to the competition parameter. $\frac{\partial^2 \lambda}{\partial x \partial \xi} > 0$ and $\frac{\partial^2 \lambda}{\partial x \partial C} < 0$ mean that a strategy with higher x is more sensitive to ξ and C, respectively. Since $\frac{\partial \lambda}{\partial C}$ is negative and $\frac{\partial \lambda}{\partial \xi}$ is positive, the terms in the parentheses are negative if increasing x increases the sensitivities. Increasing sensitivity is advantageous, if the covariance is "smoothing", i.e., if $COV(\xi, C)$ is positive. In this case the whole term $-\left(\frac{\partial^2 \lambda}{\partial x \partial \xi} \cdot \frac{\partial \lambda}{\partial C} + \frac{\partial^2 \lambda}{\partial x \partial C} \cdot \frac{\partial \lambda}{\partial \xi}\right) COV(\xi, C)$ becomes positive, indicating a positive shift in x characterizing the ESS. A disadvantageous negative covariance results in a negative term: x should be decreased to avoid a high variance in $\lambda(t)$.

(3b) $\frac{\partial^3 \lambda}{\partial x \partial \xi \partial C} \cdot COV(\xi, C)$: *"It is a wonderful possibility to have a weaker sensitivity to competition when the competition parameter is high."*

A good environment may lower the sensitivity to competition, and in this case $\frac{\partial^2 \lambda}{\partial \xi \partial C}$ is positive. Assuming positive covariance between ξ_t and C_t, years with high competition tend to be good years, and in good years the competition sensitivity is low. Having low sensitivity when competition is high is obviously advantageous, giving a higher average annual growth rate. Positive covariance therefore induces an increase in x if the sensitivity decreasing effect of good years is strengthened by increasing x ($\frac{\partial^3 \lambda}{\partial x \partial \xi \partial C} > 0$). The result is reversed when increasing x makes a good environment amplify the sensitivity to competition; or when covariance is negative (high competition and bad years are associated).

We can easily relate Eq. (5) to the previous density independent models of León (1983) and Schaffer (1974). If λ does not depend on density, $\frac{\partial \lambda}{\partial C} = 0$ and the terms corresponding to (1) and (2c), (2d), (3a), (3b) are zero. As we show in the Appendix, some coefficients containing $\lambda(x_e, \overline{\xi})$ emerge in the density independent case. In the density regulated model of Eq. (5) $\lambda(x_e, \overline{\xi}, \widehat{C})$ equals to 1, hence these coefficients didn't appear explicitly in Eq. (5); in a population without density regulation, however, λ is usually different from one, and that's why this complication arises. Since $\lambda(x_e, \overline{\xi})$ is a positive constant, the new coefficients don't alter the qualitative conclusions. So we get, as an analogue of Eq. (5) for exponentially growing populations,

$$\frac{\partial \Delta}{\partial x_r}\Big|_{x_r=x_e} = \overline{\left(\frac{1}{\lambda}\right)} \cdot \frac{\partial \lambda}{\partial x}\Big|_{x_e, \overline{\xi}} + \left[-\frac{1}{\lambda^2(x_e, \overline{\xi})} \cdot \frac{\partial^2 \lambda}{\partial x \partial \xi} \cdot \frac{\partial \lambda}{\partial \xi} + \frac{1}{2} \cdot \frac{1}{\lambda(x_e, \overline{\xi})} \cdot \frac{\partial^3 \lambda}{\partial x \partial \xi^2} \right] V(\xi) \quad (6)$$

For populations without age structure, λ is the sum of effective fecundity, B, and parental survival, p (Charnov and Schaffer 1973). Let x mean the reproductive effort, which increases fecundity and decreases parental survival. León (1983) assumed that one of the two terms in λ, e.g. the effective fecundity, is multiplied by an arbitrary function of some fluctuating environmental parameter. Therefore the annual growth rate is

$$\lambda(x, \xi_t) = B(x)f(\xi_t) + p(x) \quad (7)$$

and substituting it into Eq. (6) we get

$$\frac{\partial \Delta}{\partial x_r}\Big|_{x_r=x_e} = \overline{\left(\frac{1}{\lambda}\right)} \cdot \frac{\partial \lambda}{\partial x}\Big|_{x_e, \overline{\xi}} + \left[-\frac{1}{\lambda^2(x_e, \overline{\xi})} \left(\frac{\partial f}{\partial \xi}\right)^2 \cdot B + \frac{1}{2\lambda(x_e, \overline{\xi})} \left(\frac{\partial^2 f}{\partial \xi^2}\right) \right] \cdot \frac{\partial B}{\partial x} V(\xi)$$
$$(8)$$

Here the term $-\frac{1}{\lambda^2(x_e, \overline{\xi})} \left(\frac{\partial f}{\partial \xi}\right)^2 B \cdot \frac{\partial B}{\partial x} \cdot V(\xi)$, which corresponds to (2a), is always negative, while the sign of the term $\frac{1}{2\lambda(x_e, \overline{\xi})} \cdot \frac{\partial^2 f}{\partial \xi^2} \cdot \frac{\partial B}{\partial x} \cdot V(\xi)$ corresponding to (2b) depends on the shape of $f(\xi)$. If $f(\xi)$ is convex, the last term is positive, hence it acts to increase the optimal x and consequently $B(x)$, the term affected by fluctuation. León called fluctuation with convex $f(\xi)$ as "promising uncertainty." Note however, that promising uncertainty acts to increase $B(x)$ not because $\lambda(\xi)$ itself is convex, but because an increase in $B(x)$ directly implies the increase of the convexity of $\lambda(\xi)$. Similarly, a concave $f(\xi)$ acts to decrease the effective fecundity, since an increase in $B(x)$ would increase the concavity of $\lambda(\xi)$: León referred to it as "threatening uncertainty." A linear $f(\xi)$ does not modify the average annual growth rate, therefore it was called "neutral uncertainty."

Replacing $f(\xi)$ by the identity function $(f(\xi) = \xi)$, we get a model analogous to that of Schaffer (1974). For the linear identity function the last term in Eq. (8) is zero, and only the risk avoiding effect of (2a) appears. So any fluctuation results in a decrease of the effort invested into the term impinged by fluctuation, illustrating the adage, *"Don't put all your eggs in one basket!"*.

For the full analysis of the competition dependent case, quantities $(\overline{C} - \widehat{C})$, $V(C)$ and $COV(\xi, C)$ are required, which are determined by the environmental fluctuation and the established population. As an example, we develop the complete analysis of a lottery model in Box 1., where density regulation is explicitly specified.

IV. Departure from optimization: coexistence and exclusion of the rare strategy

In this section we present a pair of examples which illustrates the possibility of coexistence and exclusion of the rare strategy due to new effective resources raised by the fluctuation. A similar example was given by Armstrong and McGehee (1980).

In the previous section density fluctuation appeared as affecting long run growth rates via $\overline{N}$ and $V(N)$ only, because the higher order moments were neglected for the Taylor approximation. Now we intend to go beyond a weak fluctuation, but to keep the analytical tractability at the same time. It can be done by choosing a special form of density dependence: $\ln \lambda$ will be a linear or quadratic function of density. (Note that linear dependence corresponds to the Ricker recruitment curve, *c.f.* Yodzis 1989 p. 54.) This density dependence ensures that long run growth rates depend only on the average density and density variance for arbitrary strong fluctuation.

Consider two competing strategies with log growth rates

$$\ln \lambda_1(t) = -a(N(t) - K_1) - b(N(t) - K_1)^2 \tag{9a}$$

$$\ln \lambda_2(t) = -c(N(t) - K_2) \tag{9b}$$

where $N(t)$ is the sum of abundances of the two strategies, and K_1, K_2, a, b and c are positive constants to choose. $\ln \lambda_2(t)$ is a decreasing function of density $N(t)$ as it is generally required. To make $\ln \lambda_1(t)$ also decreasing for all $N(t) > 0$, we must choose the parameters to satisfy the condition

$$K_1 < \frac{a}{2b}. \tag{10}$$

Obviously, K_1 and K_2 are the equilibrium densities of strategy 1 and 2, respectively, in a stable environment. These strategies exploit a single limiting resource (indicated by the single density parameter N), correspondingly in stable environment they are not able to coexist: if $K_1 > K_2$ then strategy 1 excludes strategy 2, while if $K_1 < K_2$ then strategy 2 excludes strategy 1.

First version: Promoting coexistence

Assume that environmental fluctuation acts only on strategy 1, as a random multiplier of its annual growth rate λ_1. Then the log growth rates become

$$\ln \lambda_1(t) = \xi(t) - a(N(t) - K_1) - b(N(t) - K_1)^2 \tag{11a}$$

$$\ln \lambda_2(t) = -c(N(t) - K_2) \tag{11b}$$

where $\xi(t)$ is an environmentally fluctuating parameter, which constitutes an ergodic process and has zero mean ($\overline{\xi(t)} = 0$). The long-term behavior is determined by the average log growth rates

$$\overline{\ln \lambda_1(t)} = \overline{\xi(t)} - a(\overline{N} - K_1) - b\overline{(N - K_1)^2}$$
$$= -a(\overline{N} - K_1) - b(\overline{N} - K_1)^2 - bV(N) \tag{12a}$$

$$\overline{\ln \lambda_2(t)} = -c(\overline{N} - K_2) \tag{12b}$$

where $V(N) = \overline{(N - \overline{N})^2}$ is the variance of density. Strategy 2, which is not directly influenced by environmental fluctuation, is nevertheless affected through density dependence; since log growth rate of strategy 2 is a linear function of density, its average is altered only by a change in the average density.

According to the standard invasibility analysis coexistence is possible if both strategies have positive boundary growth rates in the established population of the other strategy. In a population of strategy 2 ($N = K_2$) stable density is determined by the equilibrium condition $\overline{\ln \lambda_2} = -c(\overline{N} - K_2) = 0$. The boundary growth rate of strategy 1 is positive if

$$\Delta_1 = -a(K_2 - K_1) - b(K_2 - K_1)^2 > 0 \tag{13}$$

and with inequality (10) this condition is equivalent to

$$K_2 < K_1. \tag{14}$$

In a population of strategy 1 density is fluctuating due to its environmentally fluctuating growth rate. The long term stochastic equilibrium condition for strategy 1 is

$$\overline{\ln \lambda_1} = -a(\overline{N}_1 - K_1) - b(\overline{N}_1 - K_1)^2 - bV(N) = 0. \tag{15}$$

Obviously, for $V(N) = 0$, $\overline{N} = K_1$ - this is the case of stable environment. Environmental fluctuation, however, provides $V(N) > 0$ and therefore leads to an average density

$$\overline{N}_1 < K_1 \tag{16}$$

because of inequality (10). Now strategy 2 has a positive boundary growth rate if

$$\Delta_2 = -c(\overline{N}_1 - K_2) > 0$$

which is satisfied if

$$\overline{N}_1 < K_2 \tag{17}$$

We can conclude from (14) and (17) that the two strategies are able to coexist if

$$\overline{N}_1 < K_2 < K_1 \tag{18}$$

where $\overline{N}_1$ means the average density in an established population of strategy 1.

Let us choose the parameters of strategy 1 (a, b, K_1) according to (10) and an arbitrary distribution for the environmental parameter $\xi(t)$. These determine $\overline{N}_1$ and $V(N)$ for the population of strategy 1. Clearly the result is independent of parameters of strategy 2 and satisfies condition (16). Next we can choose K_2 within the interval between $\overline{N}_1$ and K_1, which guarantees that both of the conditions in (18) are satisfied. c may be any positive number. With these parameters the two strategies are coexisting, while if K_2 is chosen out of the interval $[\overline{N}_1, K_1]$, coexistence becomes impossible and the strategy with the higher equilibrium density outcompetes the other one.

Fig. 6a illustrates how this model works. Here the curves corresponding to $\overline{\ln \lambda_1(t)} = 0$ and $\overline{\ln \lambda_2(t)} = 0$ are plotted in an $(\overline{N}, V(N))$ plot. Note that this figure is remarkably similar to the well known textbook figures explaining coexistence of two species on two

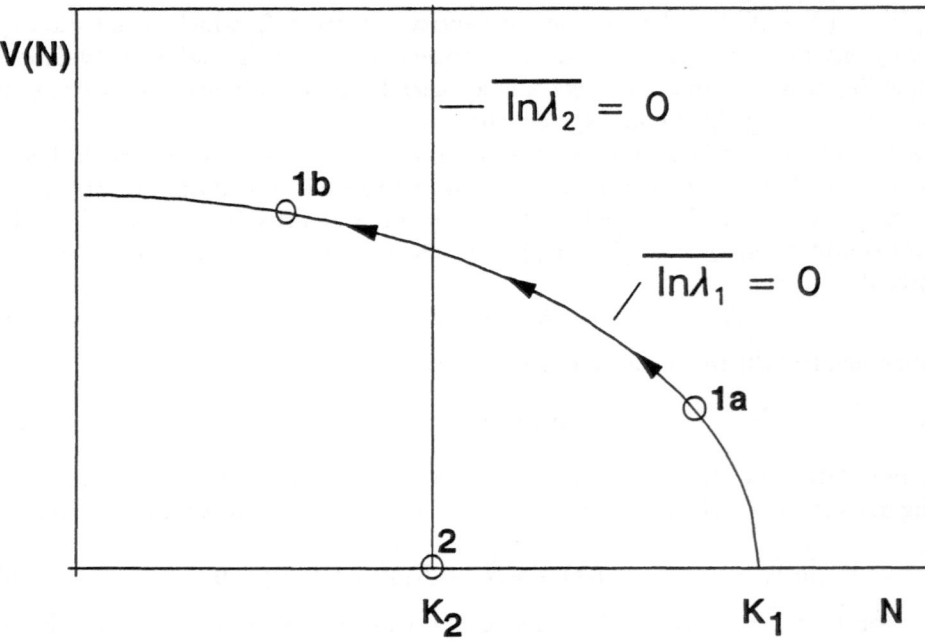

Figure 6. Average density and density variance as effective resources. Pairs $(\overline{N}, V(N))$ giving zero long run growth for strategy 1 and 2 are on a parabola and on a vertical line, respectively. Strategy 1 can increase below the parabola, while strategy 2 can increase left of the vertical line. Open circles correspond to homogeneous established populations; K_1 and K_2 are the equilibrium densities of strategy 1 and 2 in stable environment (zero density variance), respectively.

(a) Fluctuation promotes coexistence. Introducing an environmental fluctuation which acts only on strategy 1, the established population of strategy 2 is not influenced (circle 2), and being below the parabola, it can be invaded by strategy 1. Environmental fluctuation leads to a density variance, and consequently a reduced average density in an established population of strategy 1: the point corresponding to the established population of strategy 1 is shifted along the parabola. The stronger is the fluctuation, the greater is the resulting variance in density, hence the lower is the average density. A weak fluctuation corresponds to an equilibrium point of strategy 1 on the right hand side of the vertical line (1a), but a strong fluctuation makes this point move to the left hand side (1b). If the point characterizing the established population of strategy 1 is on the left hand side of the line, strategy 2 can invade the established population of strategy 1. Points 1b and 2 refer to a situation when both of the strategies can invade each other.

limiting resources (e.g. Maynard Smith 1974, Yodzis 1989), but here the axes refer to the moments of the common density distribution influencing the long run growth rates instead of the resource abundances (or the separate densities corresponding to multiple resources, cf. Section II). An established population of strategy 2 determines $\overline{N} = K_2$ and $V(N) = 0$ and can be invaded by strategy 1 if $K_2 < K_1$ irrespective of the environmental fluctuation. The position of a population of strategy 1 on this plot, however, depends on the amount of environmental fluctuation: a weak fluctuation leads to a smaller $V(N)$ than a strong one. If fluctuation is strong enough, the equilibrium

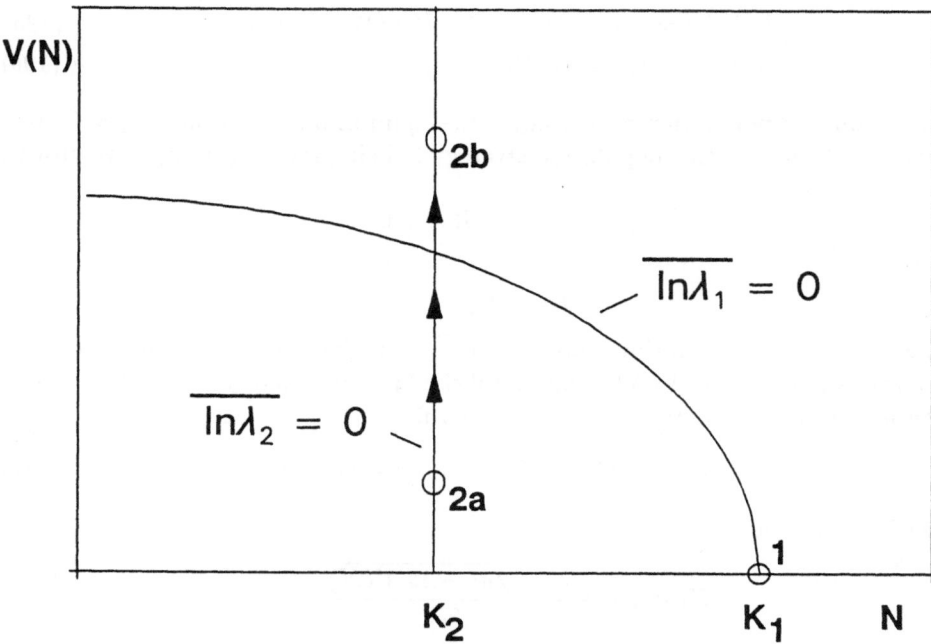

Figure 6b. Fluctuation causes the rare strategy to become extinct. Here environmental fluctuation shifts the point corresponding to the established population of strategy 2 along the vertical line. When this point remains below the parabola (2a), strategy 1 can invade strategy 2; a strong fluctuation, however, moves this point above the parabola (2b), which means that strategy 1 cannot invade strategy 2. An established population of strategy 1 has a constant density K_1, and cannot be invaded by strategy 2. Therefore points 1 and 2b describe a situation where the rare strategy is excluded.

point of strategy 1 is located on the left hand side of the line $\overline{\ln \lambda_2(t)} = 0$. In this case strategy 2 can invade the established population of strategy 1, while a population of strategy 2 remains invadable for strategy 1: the two strategies can coexist.

Replacing N by an appropriate monotonous function $f(N)$, which has not been introduced for simplicity up to now, this model can be made to satisfy the assumption of undercompensating density dependence of Chesson and Ellner (1989). In this case the boundary growth rates are proven to determine the long term evolution of a mixed population as well.

Second version: Exclusion of the rare strategy

A simple modification of the previous model leads to fluctuation-mediated disadvantage of the rare strategy; if fluctuation is strong enough, this causes the rare strategy, whichever it is, to become extinct. While in the first model version strategy 1 was affected by fluctuation, now assume that environmental fluctuation influences strategy 2 only. Therefore log growth rates are

$$\ln \lambda_1(t) = -a(N(t) - K_1) - b(N(t) - K_1)^2 \qquad (19a)$$

$$\ln \lambda_2(t) = \xi(t) - c(N(t) - K_2) \qquad (19b)$$

In an established population of strategy 1 the equilibrium condition $\overline{\ln \lambda_1} = 0$ gives stable density $N = K_1$. In this population strategy 2, being rare, has a negative growth rate if

$$\Delta_2 = -c(K_1 - K_2) < 0$$

or

$$K_2 < K_1 \qquad (20)$$

On the other hand, equilibrium condition $\overline{\ln \lambda_2} = 0$ gives an average density $\overline{N}_2 = K_2$ and some positive value of $V(N)$ in an established population of strategy 2. Strategy 1 emerging here in low frequency will be excluded if

$$\Delta_1 = -a(K_2 - K_1) - b(K_2 - K_1)^2 - bV(N) < 0 \qquad (21)$$

which is satisfied whenever

$$K_1 < K_2 + \frac{a - \sqrt{a^2 - 4b^2 V(N)}}{2b} \qquad (22)$$

Since the second term on the right hand side is positive for $V(N) > 0$, conditions for exclusion of the rare strategy

$$K_2 < K_1 < K_2 + \frac{a - \sqrt{a^2 - 4b^2 V(N)}}{2b} \qquad (23)$$

always can be satisfied by choosing K_1 not too much higher than K_2.

This model version is illustrated on Fig. 6b. The only difference in this figure compared to Fig. 6a is that environmental fluctuation acts on the position of the established population of strategy 2. If the variance in density is high enough, both of the strategies become unable to invade the other one: the rare strategy is excluded.

V. Pairwise global analysis: Who can invade whom?

Models for fluctuation mediated coexistence, like the example presented in the previous section, usually consider only two particular strategies. We can broaden this approach asking which pairs of a given strategy set can coexist in a given fluctuating environment. Recognizing all possible coexisting strategy pairs requires a standard invasibility analysis for each pair of strategies. (For practical reasons, a continuous strategy set may be approximated by an appropriate finite set of strategies.) This method, which we call *pairwise global invasibility analysis*, is a simple extension of the standard invasibility analysis introduced by Turelli (1978). Besides establishment of coexisting strategies, it provides full information about how many and which strategies are ESSs, whether the ESS can invade other populations, whether it is (they are) convergence stable, and gives some insight into the possible historical sequences of evolution in the population.

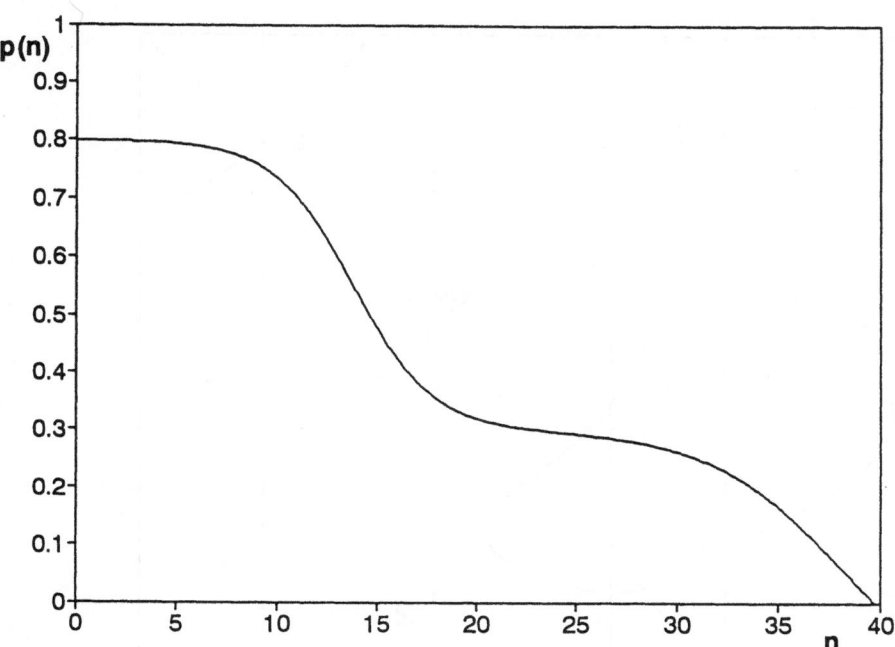

Figure 7. Pairwise invasibility analysis of the lottery model. (a) Shape of the trade-off function given in Eq. (24).

Generally the pairwise invasibility analysis can be carried out only numerically. Here we present a scheme for the numerical analysis and illustrate it with an example of the lottery model (cf. Box 1).

(1) Model definition. Give the set of the possible strategies and specify how the strategy x determines the life history parameters. Define how the fluctuating environmental parameter acts on the life history parameters and specify its distribution. Specify the density regulation: define the competition parameter and establish how the annual growth rate depends on it.

For the lottery model, we characterize strategies by the number of competing juveniles produced ($x = n$). Adult survival is traded off with fecundity, for example let the trade-off function be given by a double sigmoid curve (Fig. 7a)

$$p(n) = 0.8 - 0.5 \left[\frac{\exp(0.5n)}{10^3 + \exp(0.5n)} + \frac{\exp(0.3n)}{10^5 + exp(0.3n)} \right] \qquad (24)$$

Fluctuation acts on adult survival as a strategy-independent random factor; i.e., adult survival is $p(n)\xi_t$ in year t. For sake of simplicity, ξ_t is assumed to have only two possible values ("good years-bad years"):

$$\xi_t = \begin{cases} 1.25 & \text{with probability } 0.7826 \\ 0.1 & \text{with probability } 0.2174 \end{cases} \qquad (25)$$

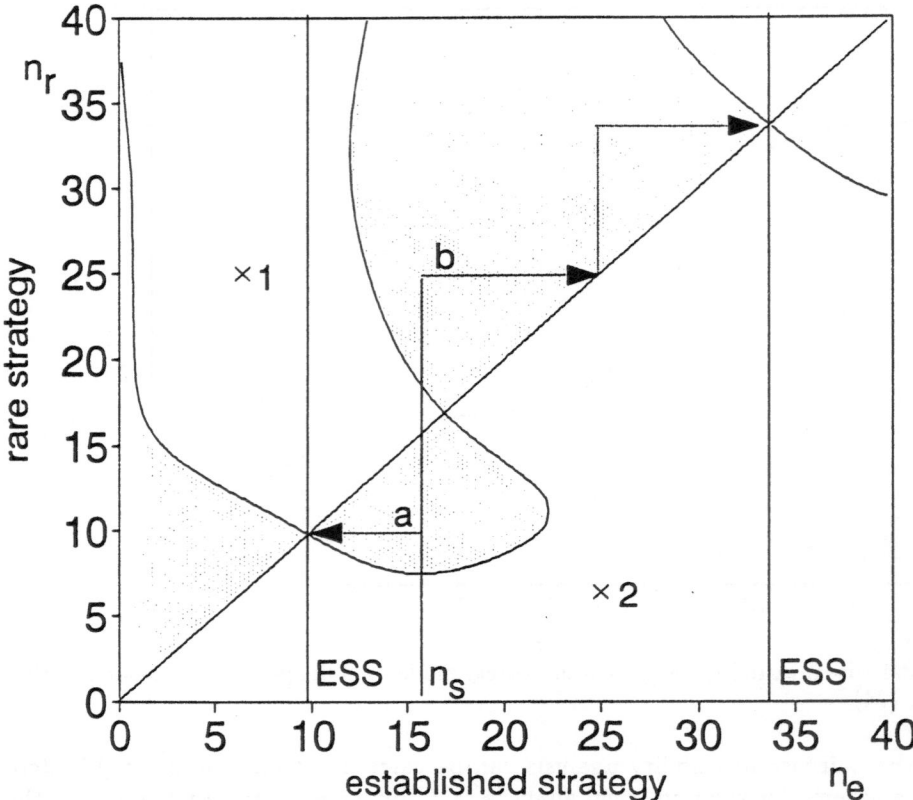

Figure 7b. The map summarizing results got from the numerical analysis of the example presented in the text. Points of the plot correspond to pairs of rare and established strategies characterized by their fecundity. Points in the dotted areas refers to pairs where the rare strategy can invade the established one; clear areas correspond to strategy pairs where the rare one cannot invade. On this map there are two ESSs, indicated by vertical lines crossing clear areas only. Points 1 and 2 refer to a case when two strategies mutually exclude each other: rare strategy $n = 25$ cannot invade $n = 6.4$ when it is established (point 1), and $n = 6.4$ cannot invade $n = 25$ when the latter is the established strategy (point 2). Two alternative historical courses are indicated by arrows a and b. The established population of strategy n_s can be invaded e.g. by the lower ESS itself; if the lower ESS appears, it will be established in the population (a). If another new mutant emerges in the population of n_s, e.g. as indicated by arrow b, the population (possibly through a series of strategies replacing each other) may reach the other ESS as well.

This distribution has an average value $\overline{\xi_t} = 1$ and a variance $V(\xi) = 0.225$. Density dependence is given by the lottery regulation described in Box 1. It is formalized by a competition parameter

$$C_t = \frac{n_e}{1 - p(n_e)\xi_t} \tag{26}$$

which acts on λ according to

$$\lambda(n, t) = \frac{n}{C_t} + p(n)\xi_t \tag{27}$$

In stable environment the optimal strategy is $n_o = 33.6829$ for the trade-off function given in Eq. (24).

(2) Choose one particular strategy x_e. Determine the joint limit distribution of (ξ_t, C_t) if this strategy is established in the population.

Generally the joint distribution can be got only from a numerical simulation of the homogeneous population of the established strategy. In some cases, like in the lottery model, the competition parameter in year t, C_t, does not depend on its previous value(s) (C_{t-1} etc.), but only on the environmental parameter in the same year, ξ_t. In these cases it is easy to find the (ξ_t, C_t) distribution analytically.

For illustration, let $n_e = 20$. From (25) and (26) the competitive parameter has two possible values, specifically for $n_e = 20$ it becomes

$$C_t = \begin{cases} 33.3129 & \text{when} \xi_t = 1.25 \\ 20.6605 & \text{when} \xi_t = 0.1 \end{cases} \tag{28}$$

(3) Choose another particular strategy x_r. Calculate the long run growth rate of x_r when joint limit distribution (ξ_t, C_t) corresponds to the established population of x_e. Long run growth rate e.g. of the rare strategy $n_r = 10$ in the established population of $n_e = 20$ can be calculated as

$$\lambda(\xi = 1.25) = \frac{n_r}{33.3129} + 1.25 p(n_r) = 1.2193$$

$$\lambda(\xi = 0.1) = \frac{n_r}{20.6605} + 0.1 p(n_r) = 0.5575$$

$$\Delta(n_r = 10, n_e = 20) = 0.7826 \cdot \ln(1.2193) + 0.2174 \cdot \ln(0.5575) = 0.0281 \tag{29}$$

Since the resulted long run growth is positive, $n_r = 10$ can invade an established population of $n_e = 20$.

(4) Repeat (2) and (3) for each pair of strategies in the set. Decide for each pair whether the rare strategy has a positive long run growth, i.e., it can invade the established one, or not.

(5) Results may be easily summarized by a map as in Fig. 7b. Every point on this map corresponds to a pair of a rare and an established strategy; we can "colour" the points according as the rare strategy can invade the established one or not. Now a lot of information can be got by looking at the map.

First, we can easily discover all the ESSs by looking for established strategies which cannot be invaded by any other strategy. In Fig. 7b two strategies ($n = 9.8335$ and $n = 33.6829$) meet this criterion: the map signs negative long run growth to all rare strategies when invading them. In this example both ESSs can invade their neighbours, but cannot invade some established strategies very different from them (ESS=9.8335 cannot invade strategies with fecundity greater than 21.7421, and ESS=33.6829 cannot invade strategies $0.4592 < n < 12.0514$).

Convergence stability can be judged from the slope of the border separating the areas where the rare strategy can and cannot invade. If this slope is negative at the

ESS, then the ESS is convergence stable and it can invade its neighbours directly as well (Fig. 8a). If the slope is positive but less than 1, the ESS cannot invade directly the other strategies similar to it population may evolve toward the ESS through a series of strategies replacing each other. Finally, the slope of the border may be greater than 1 (Fig. 8c), which implies that the ESS cannot invade its neighbours and it is convergence unstable. In the example presented in Fig. 7 the slopes at both of the ESSs are negative, hence these ESSs are convergence stable.

Pairs of strategies which can coexist or which mutually exclude each other can be identified by examining the diagonal symmetry of the map. Note that exchanging rare and established strategies ($x_r = x_1, x_e = x_2$ vs. $x_r = x_2, x_e = x_1$) we arrive at a point on the map which is diagonally symmetrical to the original one. If both of these points lie in an area indicating that the rare strategy can invade, then these two strategies can mutually invade each other, hence they can coexist according to standard invasibility analysis. The points plotted on Fig. 7b show two strategies which mutually exclude each other, since neither of them can invade the established population of the other. If one of the symmetrical points lies in a "rare can invade", but the other one lies in a "cannot invade" area, one strategy can invade and exclude the other one; hence it may be expected to spread in the population (Chesson 1982, Chesson and Ellner 1989).

Finally, the map of pairwise invasibility analysis may give some information about the historical course of evolution in the population, if (a) there are no coexisting strategies, and (b) emergence of new strategies is rare compared to the time needed for establishment. Assumptions (a) and (b) ensure that a new strategy emerges always in a homogeneous established population, hence its long run growth can be learnt from the map. Fig. 7b shows an interesting result of such an investigation. Here the population starting from a certain established strategy n_s may arrive at either of the two ESS, depending on which new strategies emerged. It means that the outcome of evolution depends not only on the initial state but also on the history of the population: which new strategies appeared at what time. Hence random mutation events may determine which of the stable endpoints will be attained.

VI. Discussion

Here we try to give some generalizations and connect the ideas which were illustrated separately by the models presented in this paper.

In Section III we determined how the ESS is shifted from the stable environment optimum in a weakly fluctuating environment. Can we regard this ESS as an optimal strategy, in the sense of Section II, in a fluctuating environment? To fulfill requirements (a) and (b) of Section II first we should investigate whether the ESS can invade other established strategies.

It is not difficult to perform a local analysis, i.e., to investigate whether the ESS can invade its neighbours, and by a reasoning similar to that of Section III we may suggest that the result is true globally as well. In Section III we considered the boundary growth rate of a rare strategy x_r in the established population of x_e, $\Delta(x_r, x_e)$, as a function of x_r for a given x_e. Now we should examine $\Delta(x_r, x_e)$ as a function of x_e, given that x_r equals the ESS. The ESS, when it is rare, can invade every established strategy x_e if

Figure 8. Convergence stability.

(a) If the border line separating "dotted" and "clear" areas has a negative slope at the ESS, the ESS is convergence stable: strategies can replace each other evolving toward the ESS (arrows). Moreover, the ESS can directly invade its neighbours (the horizontal line referring to the ESS when it is rare lies in a "dotted" area).

(b) If the slope of the border line is positive and less than 1, the ESS is convergence stable (arrows), but it cannot directly invade other strategies in its neighbourhood (the horizontal line is in a "clear" area).

(c) If the slope is greater than 1, the ESS is convergence unstable and cannot invade its neighbours.

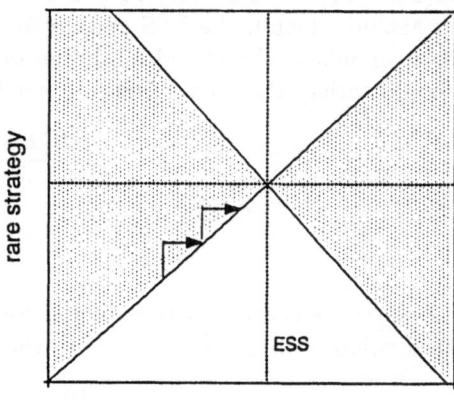
(a)

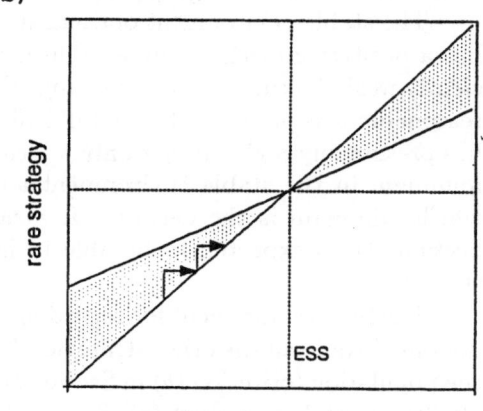
(b)

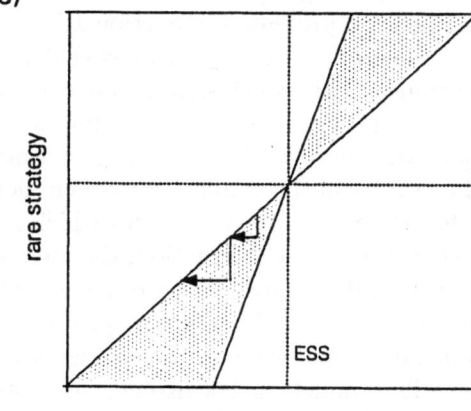
(c)

$\Delta(ESS, x_e) > 0$ for every $x_e \neq ESS$. Since $\Delta(ESS, ESS) = 0$ (no strategy can invade or exclude itself), the ESS can invade all the other established strategies if $x_e = ESS$ corresponds to the global minimum of $\Delta(ESS, x_e)$ in its variable x_e. Hence $x_e = ESS$ should satisfy the local minimum conditions

$$\frac{\partial \Delta(ESS, x_e)}{\partial x_e}|_{x_e=ESS} = 0 \qquad (30a)$$

$$\frac{\partial^2 \Delta}{\partial x_e^2} > 0 \qquad (30b)$$

Now we show that these conditions are satisfied when the ESS is the rare strategy. Derivation of $\Delta(x, x) = 0$ leads to the conclusion

$$\frac{\partial \Delta}{\partial x_r}|_{x_r=x_e} + \frac{\partial \Delta}{\partial x_e}|_{x_r=x_e} = 0 \qquad (31)$$

Since the ESS satisfies $\frac{\partial \Delta}{\partial x_r}|_{x_r=x_e} = 0$, Eq. (31) implies Eq. (30a) for $x_r = x_e = ESS$.

The stable environment optimal strategy corresponds to a minimum in x_e, since it has a positive growth in any established population except its own one. Since a sufficiently weak fluctuation cannot change the qualitative features of the shape of Δ, we can reasonably accept that (30b) is fulfilled for a weak fluctuation. Moreover, $\Delta(ESS, x_e)$ has presumably a global, not only a local minimum at $x_e = ESS$, because it is a global minimum in the stable environment and a sufficiently weak fluctuation cannot alter which minimum is the deepest. It means that the ESS, which has been explored in Section III, is expected to be able to invade the established population of any other strategy.

The final requirement for regarding the ESS as an optimal strategy is that it should not only invade all the other strategies, but also spread and finally become established in the population (criterion (b) in Section II). If we accept the results of Chesson and Ellner (1989) as valid in general (cf. Section II, $Ad(b)$), then criteria (a) and (c) imply this last one, too. Therefore we suggest that in spite of the conceptual difficulties detailed in Section II, in a *weakly* fluctuating environment there is an optimal strategy, which differs only slightly from the stable environment optimum, and can be determined by the method presented in Section III.

A corollary of this result is that a sufficiently weak fluctuation cannot lead to coexistence. This conclusion is in accordance with the message of Section IV: fluctuation mediated coexistence requires a fluctuation stronger than a minimal amount. If fluctuations were very weak, the average density in a population of strategy 1, $\overline{N}_1$ would be in the vicinity of the stable environment density, K_1. In this case there is only a very small interval $[\overline{N}_1, K_1]$ to choose K_2 which results in coexistence (cf. Eq. (18) and *Fig.* 6a). The weaker is the fluctuation, the more restrictive are the requirements for coexistence. Inversely, if two strategies are given, with some finite difference between them, we can give a finite limit for the strength of fluctuation: if the fluctuation is weaker than the limit, it cannot mediate the coexistence.

The model on coexistence presented in section IV is an apparent, very simple example for Levins' theory (Levins, 1979). According to him, fluctuation can promote coexistence because time averages of nonlinearities in the resource dependence behave as new resources: *"The number of species coexisting on resources cannot exceed the number of resources plus the number of distinct nonlinearities, or the number of effective resources is the number of original resources plus the number of distinct nonlinearities."* In our model a new effective resource, $V(N)$ arises in fluctuating environment due to the second order term in density dependence of strategy 1.

More than one resource, however, doesn't necessarily lead to stable coexistence - especially as in the second model version of Section IV. As is well known, two species can coexist when the intraspecific competition is stronger than the interspecific one. The first version of the model was figured out to meet this criterion. Here a fluctuation in density, caused by presence of strategy 1, lowers the long run growth of strategy 1, but does not influence strategy 2. Consequently here fluctuation increases the intraspecific competition for strategy 1 while it does not affect the interspecific competition between the two strategies. The increased intraspecific competition makes possible the coexistence. The second model version illustrates just the opposite situation. Here density fluctuation is connected to the presence of strategy 2 rather than strategy 1, but it causes a disadvantage for strategy 1, as before. Therefore environmental fluctuation increases the interspecific competition but does not affect the intraspecific one, resulting in the exclusion of the rare strategy. Formally an unstable coexistence point can be found, but it is surely a mathematical formality: an unstable fix point really is not a fix point in a noisy environment.

Two resources cannot maintain two species, if one of them is not limiting; i.e., if one of them being in very large abundance does not constrain the growth of any species. This is paralleled by the models in Section IV, if the fluctuation is too weak. Since $\overline{N}$ and $V(N)$ decrease the long run growth rates, they are analogous to the inverse of resource abundances. A weak fluctuation results in a small value of $V(N)$, which corresponds to a high resource abundance: a small density fluctuation cannot ensure coexistence just in the same way as a very abundant resource cannot do.

It is important to emphasize that the coexisting strategies in Section IV differed in the form of density dependence (λ was a quadratic function of density for strategy 1 while it was a linear function for strategy 2). As Metz and Godfray have proved, if $\ln \lambda$ of each strategy is a linear function of the same (generally nonlinear) decreasing function $f(N)$, there is no coexistence or rare disadvantage due to any (environmentally induced, cyclic or chaotic) fluctuation in density; the strategy minimizing the average $f(N)$ overcompetes all the other strategies (Metz and Godfray *in prep.*). Only the averages of different nonlinearities can behave as different resources.

There is not known a general simple rule to decide whether an environmental fluctuation really mediates or does not mediate coexistence. A possible source of coexistence has been investigated in a very general form by Chesson (Chesson and Huntly 1988, Chesson 1989). He characterized the competitive situation by two features: (1) the interaction between environment and competition, i.e., the quantity

$$\gamma = \frac{\partial^2 \ln \lambda}{\partial E \partial C}$$

(where E and C are some parameters describing the environment and the competition, respectively) and (2) the covariance between E and C, which was labelled by χ. Chesson introduces a reparametrization of E and C, which does not alter the product of γ and

χ. After reparametrization, he could decompose the boundary growth rate of a rare strategy i in the established population of strategy j as

$$\overline{g}_i = \Delta E - \Delta C - \gamma_j(1 - \theta b_e b_c)\chi_{jj} \tag{34}$$

where ΔE and ΔC are time-averaged differences, while the expression in parentheses usually may supposed to be positive (equation (14) in Chesson 1989). If the product of the interaction and covariance factors $(\gamma_j \cdot \chi_{jj})$ is negative, a positive term is added to the long run growth rate of the rare strategy, independently of which strategy is the rare one. Increasing the boundary growth rate of both strategies when they are rare inevitably promotes their coexistence.

In our model presented in Section IV environmental fluctuation makes coexistence possible by another mechanism. For this model $C = N$ is the (non-reparametrized) parameter describing competition, and $E = \xi$. The interaction term

$$\gamma = \frac{\partial^2 \ln \lambda}{\partial \xi \partial N} = 0$$

for both strategies, i.e., there is no interaction between competition and environment; moreover, the coexistence does not depend on any covariance between E and N. Therefore applying Chesson's formula to our model the third term is zero. Nevertheless, environmental fluctuation may lead to coexistence by an effect arising in the second term (ΔC).

The term $\Delta E = \overline{E}_i - \overline{E}_j$ cannot promote coexistence, because exchanging strategy i and strategy j ΔE always changes its sign: it cannot increase the boundary growth rate of both strategies. The situation is, however, more complicated for the term $\Delta C = \overline{C}_i - \overline{C}_j$, because quantities $\overline{C}_i$ and $\overline{C}_j$ depend on which of the strategies is the rare one. Therefore exchanging the strategies the sign of ΔC does not necessarily alter: ΔC may promote coexistence if it remains negative, while it may cause a disadvantage for the rare strategy if it is positive in both cases. As can be checked, the two versions of our model in Section IV correspond to these possibilities. On the other hand, coexistence cannot arise from ΔC if it always changes its sign by exchanging the two strategies - which is the case, for example, in the lottery model investigated by Chesson (see equation (39) in Chesson 1989). Chesson and Huntly (1989) discusses both of the mechanisms promoting coexistence through the second and the third terms.

Sections III and IV represent two traditionally different approaches to studying competition: one of them is to look for a strategy in a continuous set which ultimately outcompetes all the other strategies, under circumstances which guarantees existence and unambiguity of such a strategy; while the other one is to investigate competition between two predefined strategies under general circumstances. In Section V we tried to relate these approaches by a pairwise numerical analysis. Using this method we can discover multiple ESS, establish their convergence stability, show whether the ESS can directly invade another strategy, find coexisting strategy pairs, and explore some historical courses of evolution.

The pairwise invasibility analysis, as an extension of the standard invasibility analysis (Turelli 1978), rests on the calculation of the expected long run growth rate $\overline{\ln \lambda}$ of each possible rare strategy in each established population. It might be argued that $\overline{\ln \lambda}$

governs the evolution only if there is an infinite number of individuals (cf. using density as a continuous variable is equivalent to assuming an infinite population size). When a strategy is rare in a finite population, it is followed by just a few individuals, hence it severely suffers from demographic stochasticity (Gillespie 1974, 1977, Yoshimura and Clark 1991) – which is not considered by the present approach. Assuming an infinite (large) established population, however, we may regard a strategy as "emerged" only if there is already so many individuals of it (still in a very low frequency relative to the whole population) that the demographic stochasticity can be ignored.

It might be also objected that in a stochastic environment, "anything may happen by chance events." For example, an ESS calculated from the pairwise global invasibility analysis may be replaced by another strategy, if, just by chance, there is a run of years favoring the invading strategy. The assumption of very rare strategies emerging in very large established populations may help again, since a very rare strategy needs a long time to increase in frequency. The probability of a long run of years which gives a realized distribution of ξ_t drastically different from the theoretical one is, however, vanishingly small. Moreover, even if another strategy has succeeded to replace the ESS by chance, it is not stable evolutionarily: the further evolution of the population is expected to lead to the establishment of a convergence stable ESS, provided that it exists.

The pairwise invasibility analysis may reveal such cases where there isn't a convergence stable ESS which would be expected to be found as the (ultimately) established strategy. Here several strategies may persist in the population; the pairwise analysis can reveal all the strategy pairs which are able to coexist. Note that here we consider pure strategies only; if there are genotypes able to mix several pure strategies, the lack of a pure ESS calls for searching a mixed one. Coexistence of strategies raises very exciting questions: for example, which pairs of strategies, or coalitions involving more than two strategies are stable against the spread of other strategies? How does an evolutionarily stable coalition build up? Is there something analogous to the convergence stability? If there are several stable coalitions attainable, what are the historical events deciding which of them will result? Do evolutionary stable coalitions exist only if there isn't an ESS, or may an ESS and some coalitions be alternative outcomes of evolution? Are there such cases where neither an ESS nor an evolutionary stable coalition exists? What are the trajectories in the space of coalitions in this case?

In a pioneer paper in life history theory, Ludwig and Levin (1991) explored the coexisting pairs which are stable against any other strategy (called *evolutionary stable combinations*, ESC) in a particular model of seed dispersal. In their model the pure disperser stably coexists with an intermediate type (which is a mixed strategy itself, adopting the two pure strategies "disperse" and "not disperse" with given frequencies) in a certain range of parameters. In the former study of the same dispersal model, Cohen and Levin (1991) introduced the term of the *evolutionary compatible strategy* (ECS), which can invade every other type, but itself is not stable against invasion; hence it may be expected to coexist with another strategy, although this coexistence is not necessarily stable. Metz *et al.* (1992) also gave an example of a sex ratio model with an ECS; the simulation of their system showed convergence toward the ECS, but near the ECS the population became dimorphic. Most of the questions about the coexisting strategies are, however, open for further research.

Acknowledgements

We are grateful to Liz Pásztor, who inspired us to deal with environmental fluctuations and helped us with a lot of stimulating discussions. We thank Dan Cohen, Colin Clark, Jin Yoshimura, and an anonymous referee for their valuable comments on a previous version of this paper. Peter Chesson also commented an earlier version of the model presented in Section IV. Alberto León and Hans Metz kindly shared their unpublished work, and Colin Clark helped us with language editing. This work was supported by the National Science Foundation OTKA I/3-2220.

References

Ågren G.I. and T. Fagerström. 1984. Limiting dissimilarity in plants: Randomness prevents exclusion of species with similar competitive abilities. *Oikos* 43: 369–375.

Armstrong R.A. and R. McGehee. 1980. Competitive exclusion. *Amer. Nat.* 115: 151–170.

Bulmer M.G. 1984. Delayed germination of seeds: Cohen's model revisited. *Theor. Pop. Biol.* 26: 367–377.

Bulmer M.G. 1985. Selection for iteroparity in a variable environment. *Amer. Nat* 126: 63–71.

Charlesworth B. 1980. *Evolution in Age-structured Populations.* Cambridge University Press, Cambridge, MA.

Charnov E.L. and W.M. Schaffer. 1973. Life-history consequences of natural selection: Cole's result revisited. *Amer. Nat.* 107: 791–793.

Chesson P.L. 1982. The stabilizing effect of a random environment. *J. Math. Biol.* 15: 1–36.

Chesson P.L. 1986. Environmental variation and the coexistence of species. In: J. Diamond and T. Case (eds.), *Community Ecology*, pp. 240–256. Harper and Row, New York, NY.

Chesson P.L. 1988. Interactions between environment and competition: How fluctuations mediate coexistence and competitive exclusion. *Lecture Notes in Biomathematics* 77: 51–71.

Chesson P.L. 1989. A general model of the role of environmental variability in communities of competing species. *Lectures on Mathematics in the Life Sciences* 20: 97–123.

Chesson P.L. and S. Ellner. 1989. Invasibility and stochastic boundedness in monotonic competition models. *J. Math. Biol.* 27: 117–138.

Chesson P. and N. Huntly. 1988. Community consequences of life-history traits in a variable environment. *Ann. Zool. Fennici* 25: 5–16.

Chesson P. and N. Huntly. 1989. Short-term instabilities and long-term community dynamics. *Trends in Ecology and Evolution* 4: 293–298.

Chesson P.L. and R.R. Warner. 1981. Environmental variability promotes coexistence in lottery competitive systems. *Amer. Nat.* 117: 923–943.

Christiansen F.B. 1991. On conditions for evolutionary stability for a continuously varying character. *Amer. Nat.* 138: 37–50.

Cohen D. 1966. Optimizing reproduction in a randomly varying environment. *J. Theor. Biol.* 12: 119–129.

Cohen D. 1967. Optimizing reproduction in a randomly varying environment when a correlation may exist between the conditions at the time a choice has to be made and the subsequent outcome. *J. Theor. Biol.* 16: 1–14.

Cohen D. and S.A. Levin. 1991. Dispersal in patchy environments: The effects of temporal and spatial structure. *Theor. Pop. Biol.* 39: 63–99.

Cooper W.S. and R.H. Kaplan. 1982. Adaptive "coin-flipping": a decision-theoretic examination of natural selection for random individual variation. *J. Theor. Biol.* 94: 135–151.

Ellner S. 1985a. ESS germination strategies in randomly varying environments I. Logistic-type models. *Theor. Pop. Biol.* 28: 50–79.

Ellner S. 1985b. ESS germination strategies in randomly varying environments II. Reciprocal yield-law models. *Theor. Pop. Biol.* 28: 80–116.

Ellner S. 1987. Alternate plant life history strategies and coexistence in randomly varying environments. *Vegetatio* 69: 199–208.

Ellner S. 1989. Convergence to stationary distributions in two-species stochastic competition models. *J. Math. Biol.* 27: 451–462.

Eshel I. 1983. Evolutionary and continuous stability. *J. Theor. Biol.* 103: 99–111.

Eshel I. and U. Motro. 1981. Kin selection and strong evolutionary stability of mutual help. *Theor. Pop. Biol.* 19: 420–433.

Gillespie J.H. 1974. Natural selection for within-generation variance in offspring number. *Genetics* 76: 601–606.

Gillespie J.H. 1977. Natural selection for variances in offspring numbers: A new evolutionary principle. *Amer. Nat.* 111: 1010–1014.

Goodman D. 1984. Risk spreading as an adaptive strategy in iteroparous life histories. *Theor. Pop. Biol.* 25: 1–20.

Kaplan R.H. and W.S. Cooper. 1984. The evolution of developmental plasticity in reproductive characteristics: an application of the "adaptive coin-flipping" principle. *Amer. Nat.* 123: 393–410.

Kisdi É. and G. Meszéna. Non-invasive ESS in a lottery model with fluctuating adult survival. *in prep.*

Kozlowski J. and S.C. Stearns. 1989. Hypotheses for the production of excess zygotes: Models of bet-hedging and selective abortion. *Evolution* 43: 1369–1377.

León J.A. 1983. Compensatory strategies of energy investment in uncertain environments. In: H.E. Freedman, C. Strobeck (eds.) *Population Biology, Lecture Notes in Biomathematics* 52: 85–90, Springer-Verlag, Berlin.

León J.A. 1993. Plasticity in fluctuating environments (This volume).

Lessard S. 1990. Evolutionary stability: one concept, several meanings. *Theor. Pop. Biol.* 37: 159–170.

Levin S.A., D. Cohen and A. Hastings. 1984. Dispersal strategies in patchy environments. *Theor. Pop. Biol.* 26: 165–191.

Levins R. 1968. *Evolution in changing environments.* Princeton University Press, Princeton, N.J.

Levins R. 1979. Coexistence in a variable environment. *Amer. Nat.* 114: 765–783.

Ludwig D. and S.A. Levin. 1991. Evolutionary stability of plant communities and the maintenance of multiple dispersal types. *Theor. Pop. Biol.* 40: 285–307.

Maynard Smith J. 1974. *Models in Ecology*. Cambridge University Press, Cambridge, MA.

Maynard Smith J. 1982. *Evolution and the Theory of Games*. Cambridge University Press, Cambridge, MA.

Meszéna G. Density-dependent optimization in relation to many interrelated traits. *in prep*.

Meszéna G. and E. Pásztor. 1990. Population regulation and optimal life-history strategies. In: J. Maynard Smith and G. Vida (eds.) *Organizational Constraints on the Dynamics of Evolution*. Manchester University Press, pp. 321–331.

Metz J.A.J and H.C.J. Godfray. The evolution of demographic traits in chaotically fluctuating populations. *in prep*.

Metz J.A.J., R.M. Nisbet and S.A.H. Geritz. 1992. How should we define 'fitness' for general ecological scenarios? *Trends in Ecology and Evolution* 7: 198–202.

Murphy G.I. 1968. Pattern in life history and the environment. *Amer. Nat.* 102: 391–403.

Orzack S.H. and S. Tuljapurkar. 1989. Population dynamics in variable environments VII. The demography and evolution of iteroparity. *Amer. Nat.* 133: 901–923.

Pásztor E. 1988. Unexploited dimensions of optimization life history theory. In: G. de Jong (ed.): *Population Genetics and Evolution*. Springer-Verlag, pp. 19–32.

Schaffer W.M. 1974. Optimal reproductive effort in fluctuating environments. *Amer. Nat.* 108: 783–790.

Schaffer W.M. and M.D. Gadgil. 1975. Selection for optimal life histories in plants. In: M.N. Cody and J.M. Diamond (eds.): *Ecology and Evolution of Communities*. Belknap Press, Cambridge, MA, pp. 142–157.

Taylor P.D. 1989. Evolutionary stability in one-parameter models under weak selection. *Theor. Pop. Biol.* 36: 125–143.

Temme D.H. and E.L. Charnov. 1987. Brood size adjustment in birds: Economically tracking in a temporally varying environment. *J. Theor. Biol.* 126: 137–147.

Tuljapurkar S. 1989. An uncertain life: Demography in random environments. *Theor. Pop. Biol.* 35: 227–294.

Turelli M. 1978. Does environmental variability limit niche overlap? *Proc. Natl. Acad. Sci.* 75: 5085–5089.

Vincent T.L. and J.S. Brown. 1984. Stability in an evolutionary game. *Theor. Pop. Biol.* 26: 408–427.

Yodzis P. 1989. *Introduction to Theoretical Ecology*. Harper and Row, Publishers, New York, NY.

Yoshimura J. and C.W. Clark. 1991. Individual adaptations in stochastic environments. *Evol. Ecol.* 5: 173–192.

Appendix

To derive Eq. (5), assume that the fluctuation is weak in the following sense:

- $V(\xi)$ and $V(C)$ are of order ϵ^2; it implies that $COV(\xi, C)$ is also not greater than $o(\epsilon^2)$,
- the third and higher central moments of the (ξ_t, C_t) distribution are not greater than $o(\epsilon^3)$, and
- $(\overline{C} - \widehat{C})$ is not greater than $o(\epsilon)$..

First we show that given the above assumptions, the difference between the average competition parameter and its value in the stable environment $(\overline{C} - \widehat{C})$ must be even smaller than ϵ, and in fact, it is of order ϵ^2. In the established population of strategy x_e the long run growth rate $\overline{\ln \lambda(x_e, \xi_t, C_t)}$ must equal to zero. By Taylor-expansion of $\ln \lambda(x_e, \xi_t, C_t)$ around $\overline{\xi}$ and $\widehat{C}$ and averaging afterward we get

$$\overline{\ln \lambda(x_e, \xi_t, C_t)} = \ln \lambda(x_e, \overline{\xi}, \widehat{C}) + \frac{\frac{\partial \lambda}{\partial \xi}}{\lambda(x_e, \overline{\xi}, \widehat{C})} \cdot \overline{(\xi_t - \overline{\xi})} +$$

$$+ \frac{\frac{\partial \lambda}{\partial C}}{\lambda(x_e, \overline{\xi}, \widehat{C})} \cdot \overline{(C_t - \widehat{C})} + o(\epsilon^2) = \frac{\partial \lambda}{\partial C} \cdot (\overline{C} - \widehat{C}) + o(\epsilon^2) = 0 \qquad (A1)$$

since $\lambda(x_e, \overline{\xi}, \widehat{C}) = 1$. Thus $\overline{C} - \widehat{C} = o(\epsilon^2)$.

To get $\frac{\partial \Delta}{\partial x_r}|_{x_r = x_e}$ note that

$$\frac{\partial \Delta}{\partial x_r}\Big|_{x_r = x_e} = \frac{\partial \overline{\ln \lambda}}{\partial x}\Big|_{x_e} = \overline{\frac{1}{\lambda} \cdot \frac{\partial \lambda}{\partial x}\Big|_{x_e, \xi_t, C_t}} \qquad (A2)$$

where averaging is taken on the distribution (ξ_t, C_t) corresponding to the established population of strategy x_e. Abbreviate the terms the average of which are $o(\epsilon^2)$ by $\omega(\epsilon^2)$; e.g. a term containing $(\xi_t - \overline{\xi})^2$ is $\omega(\epsilon^2)$, since its expectation, $V(\xi)$, is $o(\epsilon^2)$. Note that this procedure is necessary because we don't want to require e.g. $(\xi_t - \overline{\xi})^2$ to be $o(\epsilon^2)$ in each particular year t; instead, only the variance $V(\xi)$ is constrained. Now $\frac{1}{\lambda}$ can be approximated by a Taylor-series around $\overline{\xi}$ and $\widehat{C}$ as

$$\frac{1}{\lambda(x_e, \xi_t, C_t)} = \frac{1}{\lambda(x_e, \overline{\xi}, \widehat{C})} - \frac{\frac{\partial \lambda}{\partial \xi}}{\lambda^2(x_e, \overline{\xi}, \widehat{C})} \cdot (\xi_t - \overline{\xi}) - \frac{\frac{\partial \lambda}{\partial C}}{\lambda^2(x_e, \overline{\xi}, \widehat{C})} \cdot (C_t - \overline{C}) -$$

$$- \frac{\frac{\partial \lambda}{\partial C}}{\lambda^2(x_e, \overline{\xi}, \widehat{C})} \cdot (\overline{C} - \widehat{C}) + \omega(\epsilon^2). \qquad (A3)$$

which, using $\lambda(x_e, \overline{\xi}, \widehat{C}) = 1$, simplifies into

$$\frac{1}{\lambda(x_e, \xi_t, C_t)} = 1 - \frac{\partial \lambda}{\partial \xi} \cdot (\xi_t - \overline{\xi}) - \frac{\partial \lambda}{\partial C} \cdot (C_t - \overline{C}) - \frac{\partial \lambda}{\partial C} \cdot (\overline{C} - \widehat{C}) + \omega(\epsilon^2). \qquad (A4)$$

Similarly, we can approximate $\frac{\partial \lambda}{\partial x}$ as

$$\frac{\partial \lambda}{\partial x}\Big|_{x_e,\xi_t,C_t} = \frac{\partial \lambda}{\partial x}\Big|_{x_e,\overline{\xi},\widehat{C}} + \frac{\partial^2 \lambda}{\partial x \partial \xi} \cdot (\xi_t - \overline{\xi}) + \frac{\partial^2 \lambda}{\partial x \partial C} \cdot (C_t - \overline{C}) + \frac{\partial^2 \lambda}{\partial x \partial C} \cdot (\overline{C} - \widehat{C}) +$$

$$+\frac{1}{2} \cdot \frac{\partial^3 \lambda}{\partial x \partial \xi^2} \cdot (\xi_t - \overline{\xi})^2 + \frac{\partial^3 \lambda}{\partial x \partial \xi \partial C} \cdot (\xi_t - \overline{\xi}) \cdot [(C_t - \overline{C}) + (\overline{C} - \widehat{C})] +$$

$$+\frac{1}{2} \cdot \frac{\partial^3 \lambda}{\partial x \partial C^2} \cdot [(C_t - \overline{C})^2 + 2(C_t - \overline{C})(\overline{C} - \widehat{C}) + (\overline{C} - \widehat{C})^2] + \omega(\epsilon^3) \qquad (A5)$$

According to Eq. (A2) we have to multiply (A4) with (A5) and average afterward. It yields

$$\overline{\frac{1}{\lambda} \cdot \frac{\partial \lambda}{\partial x}}\Big|_{x_e,\xi_t,C_t} = \overline{\left(\frac{1}{\lambda}\right)} \cdot \frac{\partial \lambda}{\partial x}\Big|_{x_e,\overline{\xi},\widehat{C}} + \frac{\partial^2 \lambda}{\partial x \partial C} \cdot (\overline{C} - \widehat{C}) + \frac{1}{2} \cdot \frac{\partial^3 \lambda}{\partial x \partial \xi^2} V(\xi) +$$

$$+\frac{\partial^3 \lambda}{\partial x \partial \xi \partial C} \text{COV}(\xi, C) + \frac{1}{2} \cdot \frac{\partial^3 \lambda}{\partial x \partial C^2} V(C) - \frac{\partial \lambda}{\partial \xi} \cdot \left[\frac{\partial^2 \lambda}{\partial x \partial \xi} V(\xi) + \frac{\partial^2 \lambda}{\partial x \partial C} \text{COV}(\xi, C)\right] -$$

$$-\frac{\partial \lambda}{\partial C} \cdot \left[\frac{\partial^2 \lambda}{\partial x \partial \xi} \text{COV}(\xi, C) + \frac{\partial^2 \lambda}{\partial x \partial C} V(C)\right] + o(\epsilon^3) \qquad (A6)$$

where $o(\epsilon^3)$ comprises all the terms containing $\overline{(\xi_t - \overline{\xi})^3}$, $\overline{(\xi_t - \overline{\xi})^2(C_t - \overline{C})}$, $\overline{(\xi_t - \overline{\xi})(C_t - \overline{C})^2}$, $(C_t - \overline{C})^3$, $[V(\xi)(\overline{C} - \widehat{C})]$, $[V(C)(\overline{C} - \widehat{C})]$, $[\text{COV}(\xi, C)(\overline{C} - \widehat{C})]$ and $(\overline{C} - \widehat{C})^2$. Note that we use $(\overline{C} - \widehat{C}) = o(\epsilon^2)$, which was proved first, when we don't write down explicitly the term which contains $(\overline{C} - \widehat{C})^2$, but we involve it in $o(\epsilon^3)$. Omitting $o(\epsilon^3)$, Eq. (A6) is equivalent to Eq. (5).

If the population has no density regulation, but grows exponentially, $\lambda(x_e, \overline{\xi})$ does not equal to 1. Hence we have to go back to Eq. (A3) to approximate $\frac{1}{\lambda}$; using $\frac{\partial \lambda}{\partial C} = 0$ we get

$$\frac{1}{\lambda(x_e, \xi_t)} = \frac{1}{\lambda(x_e, \overline{\xi})} - \frac{\frac{\partial \lambda}{\partial \xi}}{\lambda^2(x_e, \overline{\xi})} \cdot (\xi_t - \overline{\xi}) + \omega(\epsilon^2). \qquad (A7)$$

Eq. (A5) simplifies into

$$\frac{\partial \lambda}{\partial x}\Big|_{x_e,\xi_t} = \frac{\partial \lambda}{\partial x}\Big|_{x_e,\overline{\xi}} + \frac{\partial^2 \lambda}{\partial x \partial \xi} \cdot (\xi_t - \overline{\xi}) + \frac{1}{2} \cdot \frac{\partial^3 \lambda}{\partial x \partial \xi^2} \cdot (\xi_t - \overline{\xi})^2 + \omega(\epsilon^3). \qquad (A8)$$

Multiplying (A7) with (A8) and averaging on the distribution of ξ_t yields

$$\overline{\frac{1}{\lambda} \cdot \frac{\partial \lambda}{\partial x}}\Big|_{x_e,\xi_t} = \overline{\left(\frac{1}{\lambda}\right)} \cdot \frac{\partial \lambda}{\partial x}\Big|_{x_e,\overline{\xi}} + \frac{1}{2} \cdot \frac{1}{\lambda(x_e, \overline{\xi})} \cdot \frac{\partial^3 \lambda}{\partial x \partial \xi^2} V(\xi) - \frac{1}{\lambda^2(x_e, \overline{\xi})} \cdot \frac{\partial^2 \lambda}{\partial x \partial \xi} \cdot \frac{\partial \lambda}{\partial \xi} V(\xi) + o(\epsilon^3)$$
$$(A9)$$

where $o(\epsilon^3)$ contains the third and higher order central moments of the ξ_t distribution. Omitting $o(\epsilon^3)$, Eq. (A9) is equivalent to Eq. (6).

LIFE HISTORY EVOLUTION AND POPULATION DYNAMICS

IN VARIABLE ENVIRONMENTS:

SOME INSIGHTS FROM STOCHASTIC DEMOGRAPHY*

Steven Hecht Orzack

Department of Ecology and Evolution
The University of Chicago
1101 East 57th Street
Chicago, Illinois 60637

What are the evolutionary consequences of environmental fluctuations? In this paper, I present results from the theory of stochastic demography which provide a *partial* answer to this central question in the evolutionary analysis of life histories.

The "problem" of life history evolution

Despite the large amount of work initiated by Cole's famous 1954 paper on the general problem of life history evolution, progress in this area has been hindered by the lack of a dynamical framework that yields insights about the evolution of life histories given information on life history structure (such as the ages of first and last reproduction), environmental variation, and variation and covariation of components of the life history. Such an analytical framework would, for example, allow one to determine the evolutionary consequences of spreading reproduction over a lifetime in different ways. A framework with this capability is one byproduct of research on stochastic demography conducted by various workers in the last ten years or so (Cohen 1977, 1979a, 1979b, Tuljapurkar and Orzack 1980, Tuljapurkar 1982a, 1982b, 1989, Roerdink 1988, Orzack and Tuljapurkar 1989). In these works growth of a population in a temporally-variable environment is modelled as a random-matrix product, i.e.,

$$\mathbf{N}_{t+1} = \mathbf{X}_{t+1}\mathbf{N}_t = \mathbf{X}_{t+1}\mathbf{X}_t \ldots \mathbf{X}_1\mathbf{N}_0 \tag{1}$$

---

* This paper is dedicated to Monte Lloyd, on the occasion of his retirement from the University of Chicago.

where $\mathbf{N}_t$ denotes a vector (the ith element of which represents the number of individuals of the ith life history stage at time t) and $\mathbf{X}_t$ denotes a projection matrix appearing at time t. The matrices representing different environments are chosen according to a stationary stochastic process. For example, the elements of such a matrix might specify age- or stage-specific vital rates (in the form of, respectively, a Leslie or a Lefkovitch matrix) or rates of transition between diapause and nondiapause stages of a life history. Nonzero elements of the projection matrix have an average and a variance which are evaluated with respect to all environmental states. This framework has obvious and important connections with previous work in demography (e.g., Leslie 1945) and in the theory of random-matrix products (e.g., Furstenberg and Kesten 1960).

An important aspect of this analysis is that the input is the commonplace sort of data that ecologists and demographers can readily collect. In particular, one deals with, for example, survivorships and fertilities, instead of with harder-to-measure objects like reproductive "effort." This is not to say that these data are always easy to obtain or that the appropriate demographic time scale of the analysis ("the length of t") is easily determined (e.g., see Vandermeer 1975, 1978). On the other hand, the virtues of an analytical framework constructed around quantities that are readily conceived of should be apparent.

There are several important results to keep in mind in regard to Eq. (1). The first is that a population has an asymptotic growth rate

$$a = \frac{1}{t}E(\ln(N_t/N_0)) \qquad t \to \infty \tag{2}$$

where N_t is the total population size at time t (Furstenberg and Kesten 1960, Cohen 1977, Tuljapurkar and Orzack 1980). Although this is a limiting result, the realized growth rate of a population can approach a within a "short" time depending on the length and structure of the life history (Tuljapurkar and Orzack 1980).

The second result to keep in mind is that the long-run distribution of population size is lognormal (Tuljapurkar and Orzack 1980). In particular, a is the mean value of the limiting distribution of the logarithm of population size (if one disregards extinction; see below). The skewed nature of this distribution reflects the fact that the temporal average of the size of a single population and the average population size in an assemblage of replicate populations at a given time are in general unrepresentative of the typical size (see Johnson and Kotz 1970 for further information on lognormal distributions). In this sense, multiplicative growth of density-independent, structured populations in a stochastic environment is like that of scalar populations (cf., Lewontin and Cohen 1969).

A third result is that the stochastic growth rate a is relevant to the evolutionary dynamics of a simple genetic model involving life-history differences. Assume that there are two alleles, B and b, at a diploid locus. The relative ordering of the growth rates for the three genotypes determines the fate of a rare allele (say, b). For example,

$$a_{BB} < a_{Bb} > a_{bb}, \tag{3}$$

is a sufficient condition for polymorphism to be maintained when selection is weak (Tuljapurkar 1982b). a_{Bb} represents the stochastic growth rate of a monomorphic population possessing the projection matrices of Bb. (Eq. (3) has a parallel in the deterministic theory of age-structured populations in that analogous inequalities among intrinsic rates of increase determine invasion in constant environments, see Charlesworth 1980.) Analogous inequalities for the geometric means of the underlying life history components do not necessarily predict the outcome of natural selection even in a temporally-uncorrelated environment (Orzack 1985). Although these inequalities strictly determine only allelic invasion, they can predict dynamical behavior after an allele attains appreciable frequency (Orzack 1985).

Of course, an alternative sufficient condition for polymorphism is that all life histories have the same stochastic growth rate. This condition reflects the potential for a nonselective polymorphism of life history differences (see below). This condition and Eq. (3) are seen then to bound the evolutionary dynamics in that they provide a nonselective and a selective explanation, respectively, for any observed life history polymorphism. I do not claim that either has priority as an explanation for a particular life history polymorphism. Yet this agnosticism about the role of natural selection has an importance that goes beyond the stochastic framework discussed here. In particular, there is no reason to dismiss the possibility that a given life history phenotype is *not* wholly the result of natural selection. It is common to ignore this possibility perhaps because the traits involved (e.g., fertilities) cause changes in numbers of individuals and that is, of course, what evolution by natural selection is about. Nonetheless, it is reasonable to think that life history *differences* need not be caused by natural selection and equality of stochastic growth rates is a sufficient condition for this to occur. In a particular environment, natural selection may determine the value of the stochastic growth rate. Yet many different life histories may share this value. This potential for neutral differentiation is not without parallel. It also occurs for sex allocation traits, which are "obviously adaptive" in the way that life history traits are (an example is a sex ratio, see Kolman 1960). The evolutionary dynamics of sex allocation traits are almost always frequency-dependent in contrast to the frequency-independent dynamics outlined above. This frequency-dependence lends dynamical coherence to a population in such a way that any individual phenotype is neutral as long as the average phenotype for the population is not perturbed (Hines 1980, 1982). To this extent, there is no strict parallel between the neutrality of these traits and the neutrality of life histories associated with equality of stochastic growth rates. Nonetheless, the similarity between the two different kinds of traits reflects a general possibility for neutral evolution of "selectively-important" traits that should not be ignored.

The inequalities in Eq. (3) are important in that they demonstrate the relevance of a to the evolutionary dynamics of a simple genetic model. This relevance should not, however, be taken as endorsement of the view that genetic details *must* be included in an evolutionary model. It may be sufficient for understanding of life history evolution to make the same sort of assumptions about phenotypes as those that implicitly underlie evolutionarily stable strategy (ESS) models, i.e., the character is determined by a large number of weakly-linked, nonoverdominant loci with little or no epistasis or pleiotropy. This sort of genetic determination engenders evolutionary dynamics that allow the ESS to be expressed by individuals (Hines 1990). There is no a priori reason to think that similar assumptions about genetic determination should not be made about life history phenotypes. At least, the simple nature of the genetic model described above does not itself render the analytical framework unrealistic. What would render it unrealistic is the demonstration that genetic details such as overdominance, pleiotropy, and epistasis play an important role in biasing the direction of life history evolution. Evidence in regard to this point is sketchy at best.

How does a depend upon the life history and upon the environment?

These dependencies can be determined exactly for some life histories and types of environmental fluctuations (Roerdink 1988, Tuljapurkar 1986, Tuljapurkar 1990, Tuljapurkar and Istock, in press). More generally, for any suitable life history and environment (see below), Tuljapurkar (1982b) showed how to construct an approximation to a composed of terms representing the contribution of the average life history to population growth, the nonpositive contribution of one-period variances of vital rates, and the contribution of two-period temporal correlations between vital rates. This "small-noise" approximation is dynamical in that it stems from the analysis of environmental sequences (sample paths) of the stochastic process. The resulting approximations can be remarkably accurate even in highly variable environments (see Orzack and Tuljapurkar 1989).

Three points about most of the exact results and the approximation scheme should be noted. First, they relate to any life history whose demography is such that the history of the population has no effect in the limit on the steady state characteristics of the population. This ergodicity condition means, for example, that the population approaches the same limiting value of a (cf., Eq. 2) regardless of the initial age distribution. As is well known from deterministic theory (e.g., Pollard 1973) such loss of memory is assured for an age-structured life history when multiple nonperiodic bouts of reproduction occur (see Taylor 1985 for further details). The best known examples of such life histories that do not satisfy this condition are those that are strictly semelparous. More generally, the ergodicity condition can be satisfied when the structure of the life history is such that more than one class of individual contributes "newborns" to the population even though only one class is reproductive (as might occur in a monocarpic plant species that has a seed bank, see below).

The second point is that *any* biologically plausible or implausible pattern of correlations among the elements within the projection matrices can be accomodated. The resulting flexibility of analysis is unparalleled in that much previous work on evolutionary demography (despite its importance) has relied upon life histories with special structures and/or restrictive assumptions about the nature of environmental variability (e.g., see Schaffer 1974). Accordingly, the analytical framework discussed here can be used to answer questions in a way that was previously impossible. An investigator wanting to compare the stochastic growth rates of distinct life history phenotypes can now properly account for the multidimensional nature of population growth in a variable environment. Such comparisons should be part of all life history analyses. This is true even in lieu of all of the "appropriate" data. Indeed, most of the near-term applications of the theory may have this heuristic nature simply because there are few data concerning temporal variability of life history traits. Or rather, there are few data that evolutionary ecologists *think* of as relating to temporal or spatial variability of life history traits. After all, many published life tables, for example, are composed of average vital rates. The variability has been suppressed not necessarily because of a belief that it is unimportant but more likely because the tools needed to understand the demographic and evolutionary consequences of environmental variability were unavailable. A good example of this is the work of Werner and Caswell (1977) on the evolutionary demography of the teasel, *Dipsacus sylvestris*. They present stage and age projection matrices for populations located in eight distinct but contiguous fields (see also Caswell 1989). Are the differences in the projection matrices the result of sampling variation engendered by the finite size of each population, i.e., is there a single underlying projection matrix? Or do the different projection matrices accurately represent different environmental states? In general, this issue can be at least partially resolved statistically (see Bierzychudek 1982 and Caswell 1989). (In the case of Werner and Caswell's data, this analysis is not possible, Caswell, pers. comm.) However, it is not as though either significant or nonsignificant differences among projection matrices absolutely implies that one or the other explanation is correct. The outlook one chooses is, for better or for worse, partly a matter of "philosophy." But this is the central biological point and it has a very practical consequence. Any evolutionary ecologist faced with the sort of data collected by Werner and Caswell should not regard the averages of the life history components as inherently more biologically important than the rest of the moments. In particular instances, it may well be that the demographic variability apparent among samples has little ecological and evolutionary consequence, but this is a matter for analysis and therein lies the importance of the stochastic framework discussed here: one can use it to *correctly* account for the dynamical consequences of variability and accurately assess their effects.

The third point is that dynamical analysis generally results in a different estimate of the stochastic growth rate than does an analysis in which a Taylor series approximation is used to account for the effects of environmental variability on the stochastic growth rate (cf., Lacey et al. 1983). As Orzack and Tuljapurkar (1989) note, in the former analysis, the first order "shape" of the average life history modifies the contributions of variances and covariances of vital rates to stochastic growth rate. In the latter, the second order "shape" of the average life history acts in this manner. What has gone underappreciated are the qualitative and quantitative consequences of this difference (see below). Even if this difference had no appreciable numerical consequences, it is nonetheless important simply because understanding the functional anatomy of the character under investigation is as essential to evolutionary understanding in demography as it is in any other evolutionary investigation. This is just one of the reasons why the criticism by Emlen (1988) of the dynamical approach is so wrong-headed. The mistaken nature of Emlen's comments is revealed by his implication (p. 173) that the evolutionary significance of a is based upon a prior assumption of maximization (the *dynamical* genesis of a has been noted above) and by his statement (p. 173) that "...The complexity of the [approximation scheme for a] makes it all but ecologically useless." The reader may wish to judge the validity of this statement in light of the results described below (see also Orzack 1985, Lande 1987, 1988, and Orzack and Tuljapurkar 1989).

What about life history evolution?

The power of the analytical framework outlined above can be seen by its specific application to one of the most important life history problems: the temporal dispersion of a fixed amount of reproductive potential. Whether the dispersion involves iteroparity, diapause, or a seed bank, the underlying demographic "tension" between the benefits of present as opposed to future reproduction is universal. What pattern of dispersion will evolve in a variable environment?

Any meaningful evolutionary comparison of life histories must be constrained in a biologically well-motivated way. In many instances, a reasonable constraint is that all life histories have identical total amounts of reproduction. The motivation is that local evolution of life history traits may be more likely the result of changes in the timing or pattern of reproduction as opposed to involving changes in, say, physiological efficiency that allow the total amount of reproduction to increase. In the specific context of age-structured populations, for example, one can impose this constraint by setting $\Sigma \ell_i m_i (= \Sigma \phi_i$, where ℓ_i and m_i refer, respectively to average values of the survivorship to age i and the fertility at age i) equal to a constant for all life histories compared. Given this particular focus, one can then use the analytical framework outlined above to provide a specific answer to the question posed above, i.e., one can determine how environmental variability affects the stochastic growth of life histories that differ, say, in their degree of iteroparity or even in being semelparous as opposed to being iteroparous.

In particular, using the analytical framework described in Tuljapurkar (1982), it is straightforward to show that

$$a \approx \ln \lambda_0 - \frac{1}{2} \left[\sum_{\alpha}^{\omega} \left(\frac{\partial \ln \lambda_0}{\partial \phi_i} \right)^2 \sigma_{\phi_i}^2 + \sum_{\alpha}^{\omega} \sum_{\alpha}^{\omega} \left(\frac{\partial \ln \lambda_0}{\partial \phi_i} \right) \left(\frac{\partial \ln \lambda_0}{\partial \phi_j} \right) \text{cov} \left(\phi_i, \phi_j \right) \right] \quad i \neq j$$

$$(4)$$

in a temporally uncorrelated environment. α and ω are the ages of first and last reproduction, ϕ_i and $\sigma_{\phi_i}^2$ are the mean and variance of the net reproduction at age i in the life history (see above), cov (ϕ_i, ϕ_j) is the covariance of the fluctuations of net reproduction at age i and j, and λ_0 is the dominant eigenvalue associated with the average projection matrix. The values of the $\partial \ln \lambda_0 / \partial \phi_i$ are easy to calculate from the eigenvalue sensitivities associated with this matrix (which are themselves simply calculated, see Caswell 1978). (An equation with the form of Eq. (4) applies to any life history that satisfies the ergodicity condition (see above) if the sums are over all variable elements in the life history and the partial derivatives are evaluated accordingly.)

As noted above, it is of evolutionary interest to know what environmental conditions allow a given set of life histories to be selectively neutral with respect to one another. In the present context, selective neutrality is defined as equality of stochastic growth rates. The set of life histories with equivalent values of a comprise an *indifference*

Table 1. A set of life histories that share a constraint on the total amount of average net reproduction over the lifetime. ($\Sigma\phi_i$ is a constant.)

Life History	ϕ_i	α	ω	$\ln\lambda_0$	T_0	D
1	1.01	1	1	0.0099	1.000	1.000
2	0.505	1	2	0.0066	1.498	0.500
3	0.2525	1	4	0.0040	2.495	0.250
4	0.16833	1	6	0.0028	3.492	0.167
5	0.12625	1	8	0.0022	4.488	0.125
6	0.101	1	10	0.0018	5.485	0.100
7	0.12625	3	10	0.0015	6.492	0.125
8	0.16833	5	10	0.0013	7.496	0.167
9	0.2525	7	10	0.0011	8.499	0.250
10	0.505	9	10	0.0010	9.500	0.500

ϕ_i is the average net reproduction at age i and is constant from the first (α) to the last age (ω) of reproduction. λ_0 and T_0 are the asymptotic growth rate and mean generation length of the average life history in a constant environment. $D(=\Sigma(\phi_i\lambda_0^{-i})^2)$ measures the degree of iteroparity of the life history. For example, when $\lambda_0 = 1.0$, D has a minimum value of $1/\omega$ when reproduction is spread evenly over all ages and a maximum value of 1.0 when the life history if semelparous. (Values shown are rounded.)

curve. Indifference curves for a set of age-structured life histories (shown in Table 1) in a temporally uncorrelated environment are shown in Figure 1. In this example, I have assumed that coefficients of variation for the ϕ_i are constant with age and that there is an age-invariant correlation among the vital rate fluctuations at a given time. Each curve is associated with a different value of the correlation ρ. For all of the curves, $a = 0.0$, i.e., the life history is assumed to be in ecological "balance." The dynamical implications of this weak kind of equilibrium are of special ecological interest and will be discussed below.

The additional assumptions about the coefficients of variation and about ρ considerably simplify the expression for a (see Orzack and Tuljapurkar 1989). So, for example, if $\rho = 1.0$,

$$a \approx \ln\lambda_0 - \frac{1}{2}\left[\frac{C_\phi^2}{T_0^2}\right] \tag{5}$$

where C_ϕ is the age-invariant coefficient of variation and T_0 is the mean generation length of the average life history. The latter quantity (defined by Leslie 1966) increases with reproductive delay and plays a special role in the evolutionary responses of life histories to variable environments (see below).

The indifference curves in Figure 1 have several biologically important features. The first is that they implicitly relate to life histories that have *different* environmental

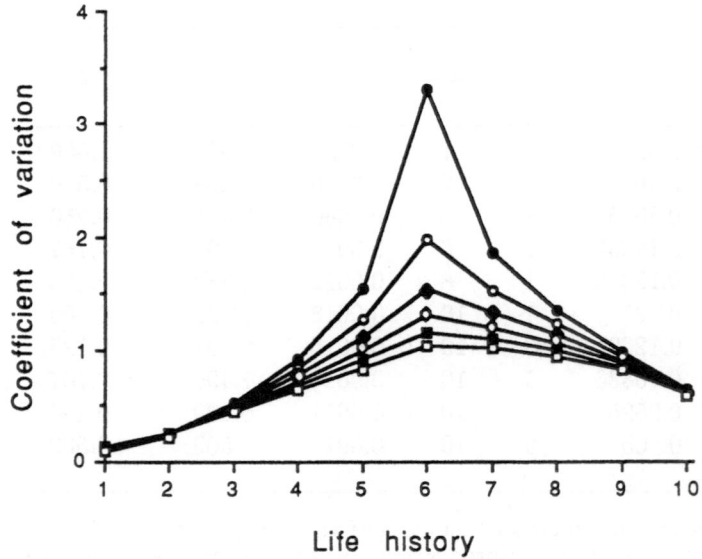

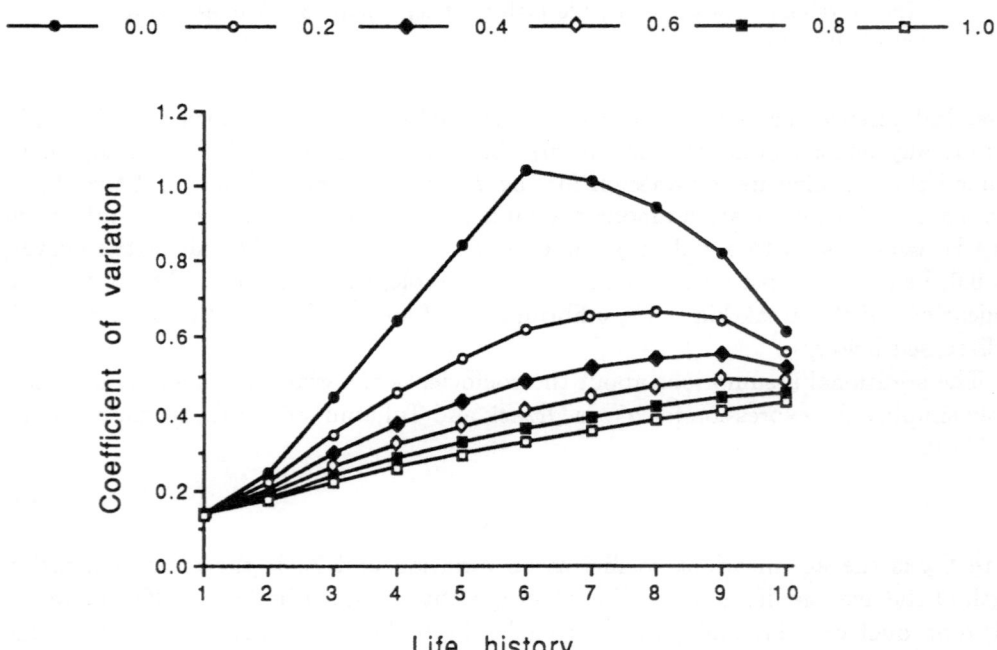

Figure 1. Indifference curves for the life histories in Table 1 for a range of values of ρ. $a = 0.0$ for all curves. Values of the coefficient of variation of the ϕ_i are derived from Eq. (4).

sensitivities. One imagines that there is an extrinsic amount of environmental variability that is experienced to different degrees by the different life histories. (Genetic variation for environmental sensitivity is described by Jinks and Pooni 1988.)

The second is that the specific form of the life history and the magnitude of the stochastic growth rate have little qualitative effect on the shape of the indifference curves. So, for example, a set of indifference curves relating to life histories with values of net reproduction that decline or peak with age are very similar to those shown here in which net reproduction is constant with age (see Figures 3 and 4 of Orzack and Tuljapurkar 1989). In addition, the general shape of the indifference curves does not change if the stochastic growth rate is larger. This can be seen in Figures 1, 2, 3, 5, and 6 of Orzack and Tuljapurkar (1989) where indifference curves associated with a range of stochastic growth rates are plotted. Similarly, changing the age of last reproduction (ω) or changing the total amount of reproduction ($\Sigma\phi_i$) has little effect on the evolutionary conclusions (see Orzack and Tuljapurkar 1989 for indifference curves associated with a larger value of $\Sigma\phi_i$).

The third point is that the shape of indifference curves does depend upon the magnitude and sign of ρ. Positive covariance of fluctuations in vital rates "magnifies" their negative effect on a. As a result, for example, the coefficient of variation associated with $a = 0.0$ for life history 6 is ≈ 3.3 when $\rho = -0.10$, whereas it is ≈ 0.30 when $\rho = 1.0$ (see Figure 1), i.e., a smaller level of environmental variation has a greater effect because the vital rate fluctuations covary positively. Accordingly, all other things being equal, natural selection favors diminished environmental sensitivity and/or life histories with negatively covarying vital rates. The character of this selective force in a temporally uncorrelated environment (towards homeostasis) differs markedly from the possible character of selection when environmental fluctuations are autocorrelated. In this situation, natural selection can favor more or less homeostasis of vital rates depending upon the identity of the vital rate that fluctuates, the form of the life history, and the sign of the autocorrelation (Orzack 1985).

The final point is that changes in life history structure need not engender changes in the relationship between environmental variability and growth rate. For example, when $\rho = 0.0$, life histories 4 and 10 have similar C values (see Figure 1) despite having very different reproductive schedules (see Table 1).

Indifference curves have implications for the comparative analysis of life history evolution. Even marked differentiation of life histories may be caused by random fixation resulting from genetic drift. It is not appropriate to assume that life history differentiation necessarily reflects adaptation to different environments (see also Schaffer 1974). On the other hand, it is not as though one need conclude on the basis of this analysis that particular cases of, say, intraspecific differentiation of life histories are the result of a nonselective process. This analysis simply indicates the degree to which very different life histories can be neutral.

As noted above, Lacey et al. (1983) suggested that the contributions of the means and variances of the underlying life history traits to the stochastic growth rate might be properly accounted for with a Taylor series, the standard way of expressing the expectation of a random variable (growth rate) as a function of the means and variances of the underlying random variables (vital rates) (e.g., see Kendall and Stuart 1977, pp. 246–247). (A Taylor series can also be used to express the variance of stochastic growth rate as a function of the underlying demographic variability.) Such an approach is based upon the prior assumption that the stochastic growth rate is equal to the geometric mean of the dominant eigenvalue of the projection matrices. This is true for a scalar life history and for some age-structured life histories with special constraints (see Tuljapurkar 1986). For most life histories, however, the stochastic growth and the geometric mean of the eigenvalues are not equal (Cohen 1977). Nonetheless, Cohen's calculations show that the geometric mean approximation can result in estimates that are close to the true stochastic growth rate. Accordingly, his examples leave unresolved whether such an approximation is "good enough" in many instances for making ecological and evolutionary conclusions about life history evolution in variable environments. That even this is not true can be seen for the following reasons. The first (Tuljapurkar 1989) is that such an approach cannot account for the dynamical consequences of temporal autocorrelation of the environmental states since only environmental frequencies enter into a geometric mean. Yet temporal autocorrelation can have appreciable effects on population and evolutionary dynamics (Orzack 1985). Even in the absence of temporal autocorrelation, it is possible for a geometric mean estimate of the stochastic growth rate to be negative when the stochastic growth rate is positive (Tuljapurkar 1989). This distinction has qualitative implications. For example, the former approach would predict certain population extinction whereas the dynamical approach would predict that the population can survive (see below). Another qualitative distinction between the two approaches that has important evolutionary implications involves the functional anatomy of life histories. Truncation of the Taylor series after the second-order terms results in the small-noise approximation,

$$ a \approx \ln \lambda_0 - \frac{1}{2} \left[\sum_{\alpha}^{\omega} \frac{\partial^2 \ln \lambda_0}{\partial \phi_i^2} \sigma_{\phi_i}^2 + \sum_{\alpha}^{\omega} \sum_{\alpha}^{\omega} \frac{\partial^2 \ln \lambda_0}{\partial \phi_i \phi_j} \operatorname{cov}(\phi_i, \phi_j) \right] \quad i \neq j \qquad (6) $$

This equation involves a further approximation in that $\ln \lambda_0$ (the logarithm of the dominant eigenvalue of the average projection matrix) replaces the average term in the Taylor series approximation (the logarithm of the average eigenvalue for the set of projection matrices). This substitution allows one to compare approximations to the stochastic growth rate that differ only in the way that variances and covariances of vital rates contribute. As shown in Eq. (4), the dynamically-based approximation contains $\partial \ln \lambda_0 / \partial \phi_i$. These sensitivities must be positive. As a result, variances and positive covariances contribute negatively to the stochastic growth rate whereas negative covariances contribute positively. In contrast, the sensitivities, $\partial^2 \ln \lambda_0 / \partial \phi_i^2$ and $\partial^2 \ln \lambda_0 / \partial \phi_i \partial \phi_j$, in Eq. (6) can be positive or negative (see Table 2). Consequently, variances and positive covariances among the ϕ_i could be mistakenly regarded as contributing positively to the stochastic growth rate. Similarly, negative covariances could be mistakenly regarded as contributing negatively to the stochastic growth rate. Caswell's (1989, p. 225) implication that the dynamical approach and the Taylor series approach are qualitatively "quite similar" overlooks this important biological distinction between the two different kinds of coefficients.

Table 2. The set of $\dfrac{\partial^2 \ln \lambda_0}{\partial \phi_i \partial \phi_j}$ for life history 6

					j					
	1	2	3	4	5	6	7	8	9	10
1	0.199	0.165	0.132	0.099	0.066	0.033	0.000	−0.033	−0.066	−0.099
2		0.132	0.099	0.066	0.033	0.000	−0.033	−0.066	−0.099	−0.132
3			0.066	0.033	0.000	−0.033	−0.066	−0.099	−0.132	−0.163
4				0.000	−0.033	−0.066	−0.099	−0.132	−0.163	−0.195
5					−0.066	−0.099	−0.132	−0.163	−0.195	−0.227
i 6						−0.132	−0.163	−0.195	−0.227	−0.259
7							−0.195	−0.227	−0.259	−0.291
8								−0.259	−0.291	−0.323
9									−0.323	−0.354
10										−0.386

ϕ_i (= 0.101) is the average net reproduction at age i and is constant from the first ($\alpha = 1$) to the last ($\omega = 10$) age of reproduction (see Table 1).

One might still ask whether the estimates of stochastic growth rates arising from Eq. (6) generally match those arising from Eq. (4). As noted, the estimates match exactly for a scalar life history, but for other life histories there is no necessary good quantitative correspondance between the estimates. This can be seen in Figure 2 in which Taylor series estimates of stochastic growth rates are plotted against "true" stochastic growth rates. The Taylor series estimates are always less than the predictions of Eq. (4) for all life histories. Of most importance are that the discrepancy depends on the "true" stochastic growth rate as well as on the life history and that the ordering of the Taylor series estimates changes with stochastic growth rate. So, for example, when the stochastic growth rate is 0.0, the Taylor series estimate for life history 3 is greater than that for life history 4. The reverse is true when the stochastic growth rate is 0.002. These dependencies imply that no simple scaling of the Taylor series estimates will render them biologically useful.

Taylor series estimates of the stochastic growth rate can be used to construct indifference curves (see Figure 3).The smaller range of coefficients of variation (as compared with Figure 1) indicates that the *net* effect of using the Taylor series approximation is to overestimate the negative effects of environmental variability on the stochastic growth rate. More importantly, the Taylor series approach does not reveal an important qualitative feature of indifference curves based on Eq. (4), i.e., that a long prereproductive period more than compensates for a decrease in average growth because it discounts the deleterious consequences of environmental variability. As shown in Eq. (5) for the case of $\rho = 1.0$, the square of the mean generation length (which reflects reproductive

S.H. Orzack

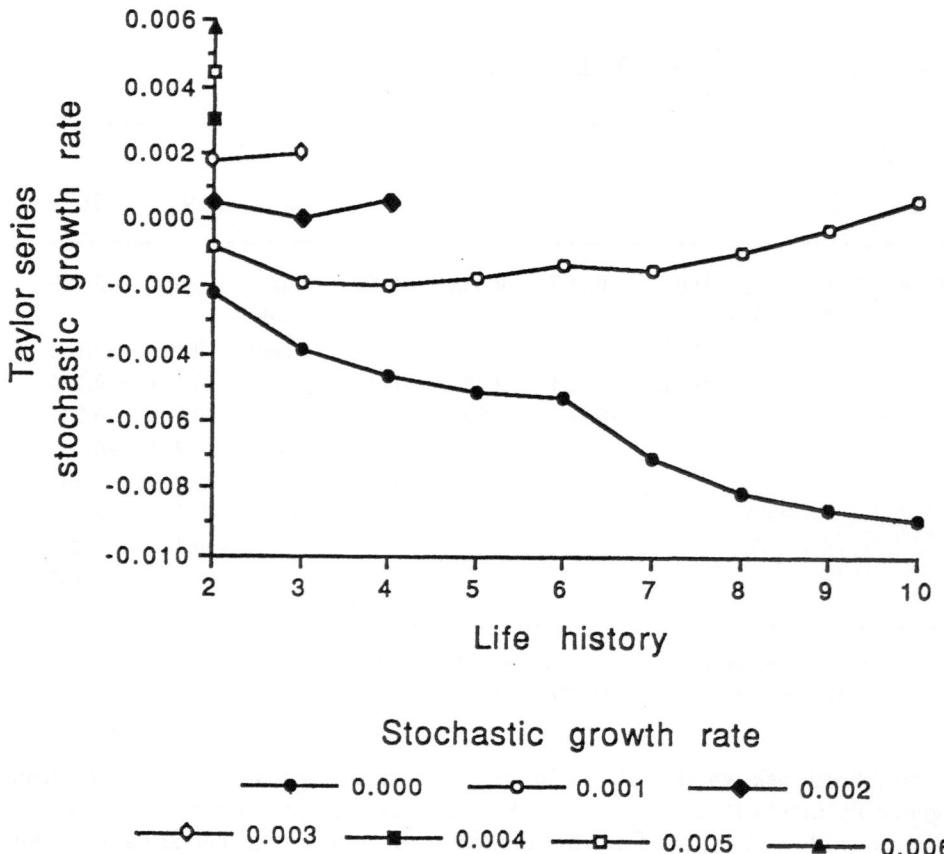

Figure 2. Taylor series stochastic growth rates (derived from Eq. (6)) plotted against life history (see Table 1) for a range of "true" stochastic growth rates. $\rho = 0.0$. Each Taylor series estimate is based upon the coefficient of variation associated with the true stochastic growth rate (derived from Eq. (4)).

delay) discounts environmental variability. As a result, the dynamically-based indifference curves increase as reproductive delay increases, i.e., life history 10, for example, can maintain a given stochastic growth rate in the face of a higher amount of environmental variability than can life history 1 (see Figure 1). In contrast, as shown in Figure 3, the Taylor series indifference curve is almost flat, i.e., life history 10 is incorrectly predicted to be no better than life history 1 in withstanding the deleterious effects of environmental variability.

For all of these reasons, it is clear that a geometric mean, Taylor series approach is not sufficient for qualitative and quantitative understanding of the population dynamics of structured life histories in variable environments.

Additional evolutionary insights can be gained by examination of the relationship between the stochastic growth rate and the amount of environmental variability (see Figure 4).

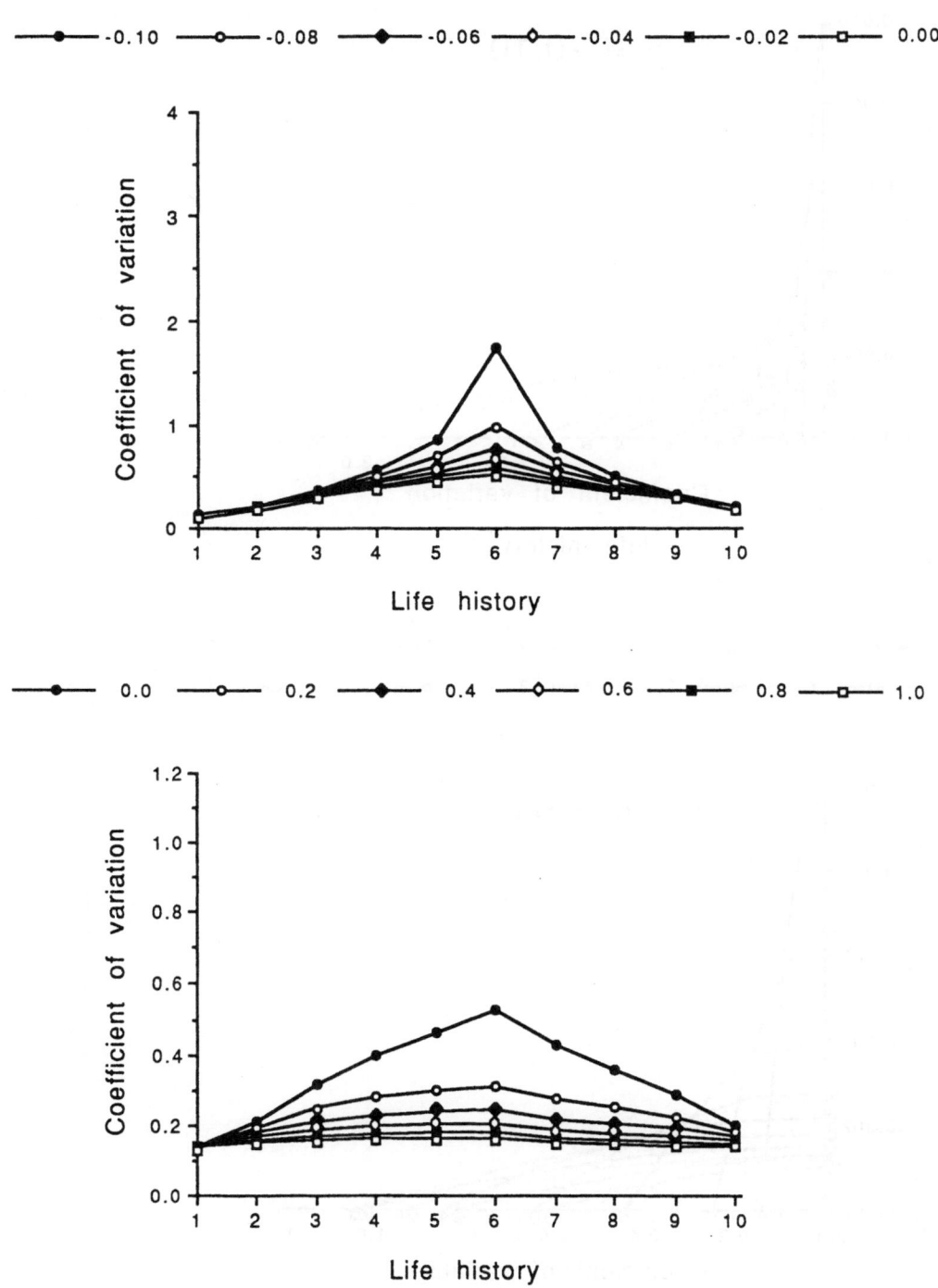

Figure 3. Taylor series indifference curves for the life histories in Table 1 for a range of values of ρ. $a = 0.0$ for all curves. Values of the coefficent of variation are derived from Eq. (6).

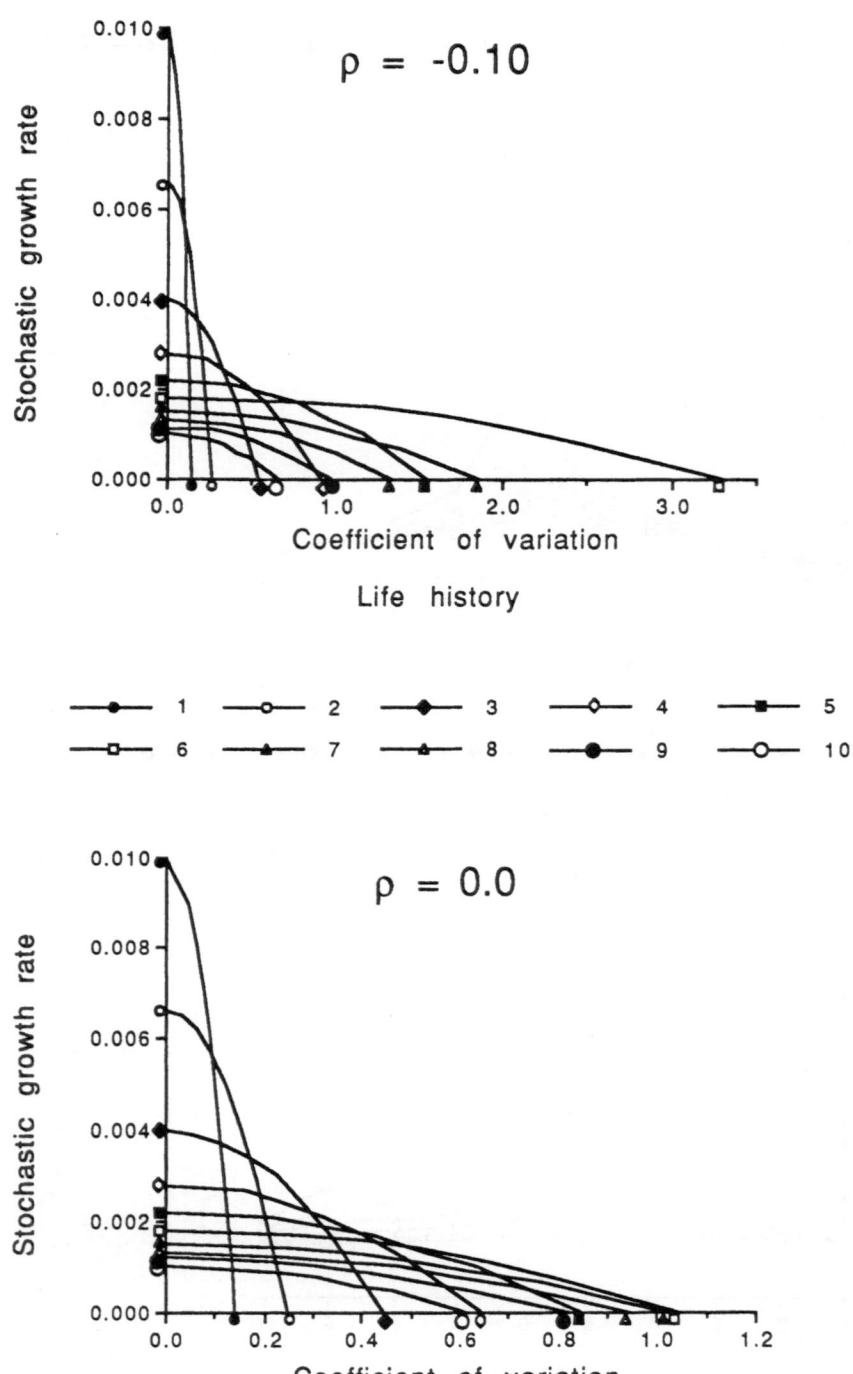

Figure 4a. The relationship between coefficient of variation and stochastic growth rate for the life histories in Table 1 for a range of values of ρ. The curves are derived from Eq. (4).

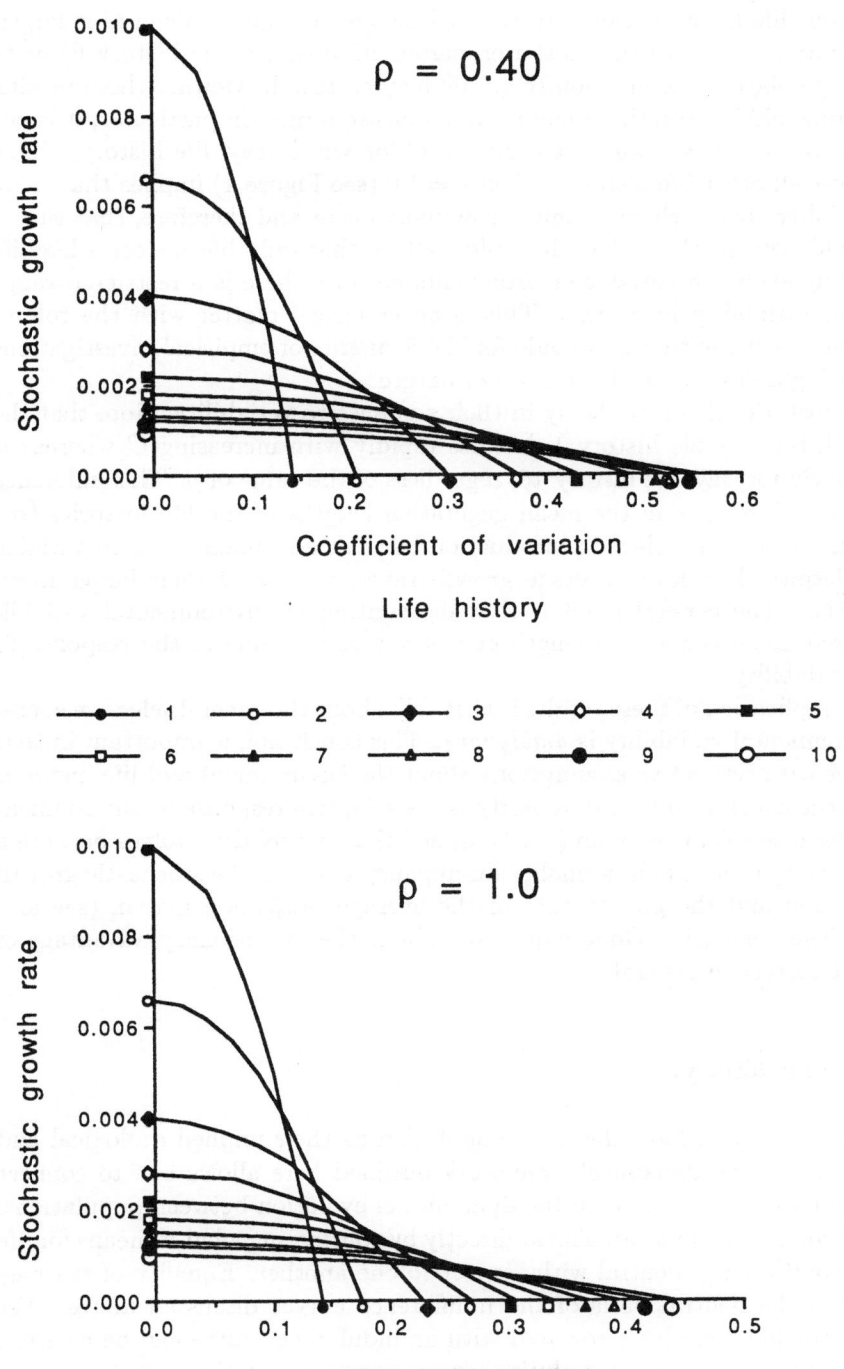

Figure 4b.

When C is *relatively* low, there is an advantage to life history 1, which is semelparous. This advantage holds regardless of the value of ρ. When C is intermediate, various iteroparous life histories can have the highest growth rates. When C is larger, there is an advantage either to the most iteroparous life history (life history 6) or to the one with very delayed reproduction (e.g., life history 10). In viewing these results, it is important to avoid interpreting them in probabilistic terms. In particular, it is not as though the relatively small range of values of C for which, say, life history 10 has an advantage over all other life histories when $\rho = 1.0$ (see Figure 4) implies that there is a small probability that such environments actually occur and therefore, that such a life history should only rarely evolve. It could well be that only life histories like life history 10 actually evolve in variable environments because there is a restricted range of environmental variability in nature. This is an empirical matter with the role of theory in this context being that of a guide for the focussing of empirical investigations as opposed to being a descriptor of the state of nature.

Finally, life histories differ markedly in their response to variability. Note that the stochastic growth rate for life history 1 declines rapidly with increasing C whereas it declines more slowly for, say, life history 10, regardless of the value of ρ. This difference stems from the the difference in the mean generation lengths of the life histories (see Table 1). Life histories with delayed reproduction have an advantage in more variable environments, despite their lower average growth rates, *because* of their larger mean generation lengths. The general point is that discounting of environmental variability by the squared mean generation length can play a special role in the response to environmental variability.

The overall implication of these results is that talk about the general selective consequences of environmental variability is *ambiguous*. This conclusion is important in that previous analyses with restrictive assumptions about the environment and life histories have resulted in the conclusion that iteroparity is *the* adaptive response to environmental variability. For example, Goodman (1984) argued that iteroparity evolves because a longer reproductive span results in a smaller discrepancy between the stochastic growth rate of a population and the growth rate of the average projection matrix (see also Murphy 1968). The conclusion Goodman draws about the evolutionary advantage of iteroparity is not correct in general.

What is a neutral life history?

All of the previous analyses have the local population as their implied ecological and evolutionary context. The dynamical framework outlined here allows one to connect these within-population dynamics with the dynamics of evolution between populations.

This connection can be illustrated most directly by considering what it means for life histories to be evolutionarily neutral with respect to one another. Equality of stochastic growth rates is the neutral basis of the indifference curves discussed above. Yet the neutrality of the life histories associated with an indifference curve can be assessed with respect to a different metric of evolutionary success: a population's probability of extinction. As Thoday (1953) and others have stressed, a population or lineage's probability of extinction is an important measure of fitness (see also Tuljapurkar and Orzack

1980). Life histories that are neutral with respect to stochastic growth rate need *not* be neutral with respect to extinction probability. In this sense, the neutrality of indifference curves is necessarily defined only with respect to within-population dynamics. (This statement may not be true in a strict sense. Populations with the same stochastic growth rate that differ with respect to their assemblages of neutral life histories may differ with respect to extinction probability. This is not a well-explored problem from a stochastic or biological point of view).

Extinction dynamics

To see how between-population dynamics can differ from within-population dynamics, imagine that otherwise identical populations are fixed for each of the different life histories associated with an indifference curve. This might occur as a result of population subdivision or a founder event. The extinction dynamics of such populations are analytically well understood (see, e.g., Tuljapurkar and Orzack 1980 and Lande and Orzack 1988). In particular, the probability of ultimate extinction in an environment with no temporal autocorrelation is

$$1.0 \;\text{ for } a \le 0.0$$

and

$$\exp\left(\frac{-2ax_0}{\sigma^2}\right) \;\text{ for } a > 0.0. \tag{7}$$

where x_0 is the logarithm of the initial population size and σ^2 is the variance of stochastic growth rate. Lande and Orzack (1988) showed that the initial population size should be adjusted to account for among-individual differences in reproductive value. This adjustment depends upon the initial age distribution, which may be difficult to determine retrospectively although x_0 is easy to calculate given any estimate (Lande and Orzack 1988). Tuljapurkar (1982) demonstrated that $\sigma^2 \approx 2(\ln \lambda_0 - a)$. This approximation of the variance of the stochastic growth rate is quite accurate even in some highly variable environments (Roerdink 1989, Orzack, unpublished).

Four features of these analytical results deserve comment.

1. When $a > 0.0$, the distribution of extinction times is improper since the probability of ultimate extinction is less than one. Yet the *conditional* distribution of extinction times when $a > 0.0$ is the same as the distribution of extinction times when $a < 0.0$ since the former does not depend on the sign of a. This result (apparently first noted by Whitmore 1978) can be derived from the theory of conditional diffusion processes (Lande and Orzack 1988).

2. When $a \ne 0.0$, the distribution of extinction times is known as an inverse Gaussian distribution (see Chhikara and Folks 1989). The mean and variance of extinction time are

$$\frac{x_0}{|a|} \quad \text{and} \quad \frac{x_0\sigma^2}{|a|^3}. \tag{8}$$

When $a = 0.0$, the distribution of extinction times is the positive stable distribution of order $1/2$ (see Feller 1971, pp. 173–175). The mean and variance are infinite, although all extinction times are finite. (The density functions of these distributions are identical.)

In general, the distribution of extinction times is such that some populations go extinct only after a long time (e.g., see Figure 1 of Dennis et al. 1991). There are important ecological implications of this persistence (see below). The distribution of extinction times is positively skewed and the modal extinction time (always less than the mean) is

$$\frac{\sqrt{9\sigma^4 + 4x_0^2 a^2} - 3\sigma^2}{2a^2} \text{ when } a \neq 0.0$$

and   (9)

$$\frac{x_0^2}{3\sigma^2} \text{ when } a = 0.0$$

(see Johnson and Kotz 1970 and Levinton and Ginzburg 1984). An important feature of these results is the ease of estimating finite-time extinction probabilities since they are based on values of the cumulative normal probability distribution (see Eq. (11) of Lande and Orzack 1988).

3. Extinction is defined as attainment of a certain critical population size (since the total population size, ΣN_i, in Eq. (1) cannot be equal to 0.0). All of the analytical and numerical results in this paper are based upon the assumption that the critical size is one composite individual, i.e., that $\Sigma N_i = 1.0$. (Defining the extinction boundary with respect to this sum does permit populations with $\Sigma N_i > 1.0$ but all $N_i < 1.0$ to avoid extinction at time t. Most such populations are likely to go extinct soon after time t.) Any population size could be regarded as being the critical value which "guarantees" extinction. For a given value for the extinction boundary, one need only take x_0 to be the difference on the logarithmic scale between the adjusted initial population size and the critical size.

4. These results concerning extinctions stem from assuming that changes in the logarithm of population size are described by the diffusion process known as the Wiener process (see Cox and Miller 1965 for further information on this stochastic process). Changes in population size are assumed to be due to environmental fluctuations that are independent, stationary, and normally distributed. The last assumption is inaccurate in that the distribution of the logarithm of population size deviates from normality at small times (implying nonnormality of increments) although it can be "quickly" achieved at least relative to the time intervals over which extinction dynamics are usually to be examined (see Tuljapurkar and Orzack 1980). The assumption of independence among the increments is violated for any age-structured population in that age-structure engenders temporal autocorrelation among the one-period growth rates (Lande and Orzack 1988). Despite these inaccuracies, the analytical results concerning extinction dynamics predict the results of stochastic simulations to within a few percent (Lande and Orzack 1988) except for some life histories with clumped and delayed reproduction (Orzack, unpublished; see below). This is not surprising because the dynamical behavior of a diffusion process with correlated stationary increments approaches that of the Wiener process at long times relative to the largest correlation time of the process (Kuznetsov et al.

Table 3. Probabilities of ultimate extinction for the life histories in Table 1.

Life History	N_0^*	$\ln \lambda_0$	Ultimate extinction probability given $x_0 =$	
			$\ln N_0^*$	$\ln$ (constant)
1	10.000	0.0099	0.771	0.771
2	13.304	0.0066	0.632	0.665
3	15.937	0.0040	0.395	0.462
4	17.062	0.0028	0.215	0.287
5	17.686	0.0022	0.094	0.150
6	18.084	0.0018	0.028	0.058
7	15.298	0.0015	0.006	0.013
8	13.261	0.0013	< 0.001	0.001
9	11.705	0.0011	$< 1 \times 10^{-6}$	$< 1 \times 10^{-5}$
10	10.477	0.0010	$< 1 \times 10^{-8}$	$< 1 \times 10^{-8}$

The initial population size is constant ($= 10.0$) or is adjusted (N_0^*) based upon the assumption that 10 newborn individuals initiate a population. $a = 0.001$. Values are derived from Eq. (7).

1965). Another practical consequence of this convergence is that it implies that the analytical results may also be useful for describing the extinction dynamics of populations experiencing exogenously-based temporally-autocorrelated environmental fluctuations.

How does life history structure affect extinction dynamics?

The claim that life histories with identical stochastic growth rates can have different ultimate extinction probabilities follows from the dependency of σ^2 on $\ln \lambda_0$. In particular, life histories with higher average growth rates ($\ln \lambda_0$) have a higher probability of ultimate extinction. This can be seen in Table 3 where probabilities of ultimate extinction are shown for the life histories in Table 1. I have assumed that x_0 is constant across populations or that 10 newborn individuals initiate each population. (All estimates of extinction probabilities in this paper are based upon one or the other of these assumptions.) Life histories with higher average growth rates have higher extinction probabilities even when x_0 is increased to account for the reproductive value of individuals. This *positive* relationship between average growth rate and extinction probability exemplifies the importance of understanding demography at the stochastic level. This point is of special applied importance given that some ecologists and conservation biologists have referred to an increased average growth rate as necessarily decreasing a population's probability of extinction (e.g., see Diamond 1984, p. 196 and Belovsky 1987, p. 38).

Table 4. Extinction time statistics for life histories in Table 1.

Life History	$\ln N_0^*$			$\ln(\text{constant})$		
	Mean	Variance	Mode	Mean	Variance	Mode
1	2303	40853240	99	2303	40853240	99
2	2588	29178810	197	2303	25960244	156
3	2769	16514640	419	2303	13734790	292
4	2837	10472322	684	2303	8500060	460
5	2873	6967672	998	2303	5584678	667
6	2895	4683618	1351	2303	3725186	919
7	2728	2885287	1569	2303	2435592	1210
8	2585	1681196	1787	2303	1497628	1525
9	2460	829132	2006	2303	776072	1852
10	2349	214406	2216	2303	210152	2170

The initial population size is constant ($= 10.0$) or is adjusted (N_0^*) based upon the assumption that 10 newborn individuals initiate a population (see Table 3). $a = 0.001$. Times (in arbitrary units) are rounded to the nearest integer.

The relationship between life history and the extinction time distribution is shown in Table 4. Mean extinction time does not increase with average growth rate but, instead, with the degree of iteroparity when $x_0 = \ln N_0^*$. (Recall that life history 1 is the least iteroparous life history in the set and life history 6 is the most iteroparous.) In contrast, the variance of extinction time decreases with average growth rate (regardless of the value of x_0). The result is that the coefficient of variation of extinction time increases with average growth rate, implying that life history 1 has the most unpredictable extinction time.

The results in Tables 3 and 4 concern life histories that share a stochastic growth rate but differ in their extinction dynamics. They illustrate the potential evolutionary distinction between selective neutrality within populations and selective neutrality between populations.

This is, of course, not a necessary distinction in that life histories associated with an indifference curve can also be neutral with respect to ultimate extinction probability. This is always true of the life histories that share a stochastic growth rate of 0.0 – such life histories will always go extinct. More generally, life histories with, say, identical positive stochastic growth rates can also share identical ultimate extinction probabilities. One way to do this is to remove the shared constraint on the total amount of reproduction. So, for example, indifference curves defined with respect to stochastic growth rate and ultimate extinction probability are shown in Figure 5. The life histories are shown in Table 5. Note that the patterns of reproduction across ages are identical to those in Table 1 but that the total amount of weighted reproduction ($\Sigma \phi_i$) now differs across life histories within a set.

The indifference curves shown in Figure 5 have some important evolutionary implications. As in the case of life histories sharing only a stochastic growth rate, the shape of the indifference curve depends upon the correlation between vital rates. For $\rho = 0.0$, the curves differ from those in Figure 1 in that life history 6 is now just one of several life histories that can maintain a stochastic growth rate of 0.0 when the environment is highly variable ($C \geq 3.0$). In this sense, these curves again reveal that life histories with dispersed reproduction need not have an evolutionary advantage in a variable environment. It is critical to note that the trends of indifference curves should be interpreted only qualitatively when the environment is highly variable. The analytical framework discussed here (cf. Eq. (4)) is known to accurately predict stochastic growth rates when $C \leq 3.0$ (see Figure 2 of Orzack and Tuljapurkar 1989). For higher values of C, the approximation can produce quantitatively inaccurate results (Orzack, unpublished).

These analyses show that indifference curves based upon equality of stochastic growth rates and of ultimate extinction probabilities can provide insight into life history evolution in variable environments. I do not claim that these curves are as evolutionarily meaningful as those based only on equality of stochastic growth rates. The reason is that in the former case, ultimate extinction probability acts as a constraint. In the latter case, there is a constraint on the total amount of reproduction. As noted above, this second kind of constraint is biologically meaningful to the extent that individuals in the same population may often be equally physiologically efficient (perhaps as a result of natural selection) such that only the timing and pattern of reproduction can vary. This, at least, is a common starting assumption for many life history analyses. In contrast, the constraint imposed by a common extinction probability is defined with respect to the long-term history of a population. In this sense, there is not a clearly-elaborated mechanism which could impose such a constraint. Perhaps some selective process which operates at the level of species or ecosystem can do so. In any case, the indifference curves defined with respect to stochastic growth rate and extinction probability are discussed here to demonstrate how such considerations can be accounted for within the present dynamical framework. This area in the study of life history evolution deserves more exploration (see also Holgate 1967). In this regard, I note that these analyses are not exhaustive with respect to the analysis of evolutionary neutrality. One could also define indifference curves solely with respect to ultimate extinction probability by, for example, constraining the total amount of reproduction and allowing the stochastic growth rate to vary.

The neutrality of the life histories shown in Figure 5 holds true for finite times. Indeed, the nature of the constraint imposed on ultimate extinction probability implies that the cumulative extinction probabilities of the life histories are identical at any time. This identity implies, of course, that the mean, variance, and modal times of extinction are identical for all life histories associated with an indifference curve. However, this need not be true for life histories that have identical stochastic growth rates and ultimate extinction probabilities and therein lie some important ecological and evolutionary insights.

Probability of extinction

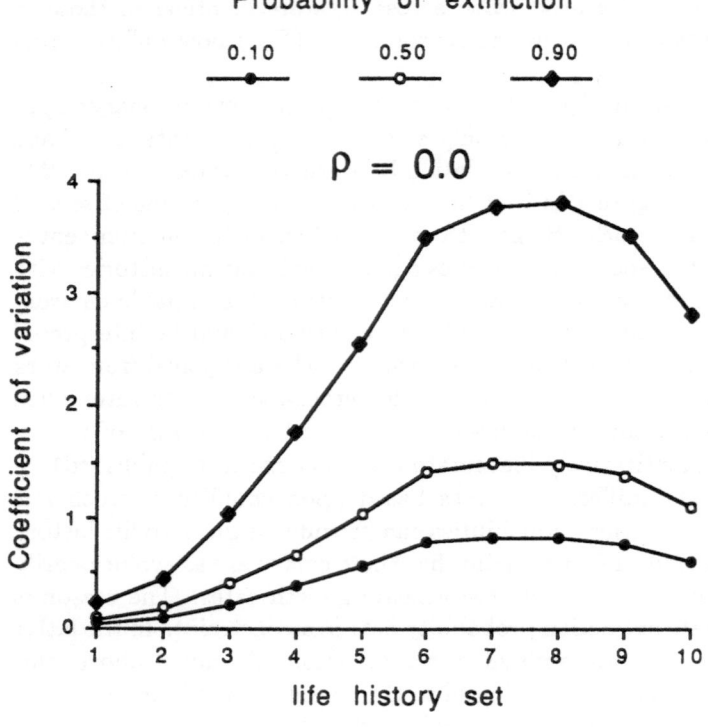

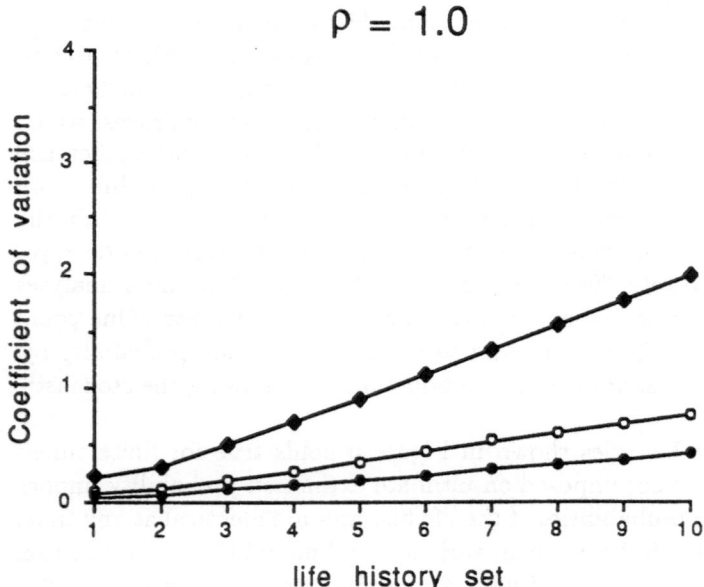

Figure 5. Indifference curves defined with respect to stochastic growth rate and ultimate extinction probability for the life histories in Table 1. $a = 0.001$ for all curves. Values of the coefficent of variation are derived from Eq. (4). x_0 is assumed constant ($= \ln(10.0)$) for all life histories. Ultimate extinction probabilities are derived from Eq. (7)).

Table 5. ϕ_i values for life histories that share a constraint on the ultimate extinction probability for a given value of $a(=0.001)$ (see Figure 5).

Life History Set	α	ω	Probability of extinction =		
			0.10	0.50	0.90
1	1	1	1.0020	1.0044	1.0236
2	1	2	0.5015	0.5032	0.5174
3	1	4	0.2512	0.2527	0.2646
4	1	6	0.1678	0.1692	0.1804
5	1	8	0.1261	0.1274	0.1383
6	1	10	0.1011	0.1024	0.1131
7	3	10	0.1266	0.1286	0.1448
8	5	10	0.1692	0.1721	0.1977
9	7	10	0.2543	0.2593	0.3035
10	9	10	0.5096	0.5210	0.6212

ϕ_i is the average net reproduction at age i and is constant from the first (α) to the last age (ω) of reproduction. x_0 is assumed to be constant ($= \ln(10.0)$).

As noted above, life histories with a stochastic growth rate of 0.0 also share an ultimate extinction probability of 1.0. Yet such life histories are not neutral in the transient sense, i.e., their extinction probabilities at finite times differ (see Figure 6). For a given life history, each curve describes the transient dynamics for any value of ρ.

Two features of these curves are important. The first is that there is no simple relationship between differences in life history structure and differences in transient extinction probabilities. Consider, for example, life histories 2, 3, 4, and 5 and life histories 7, 8, 9, and 10. These sets have the same range of differences in the number of reproductive events during life (see Table 1). Yet, after 1000 time units, for example, the life histories in the first set differ substantially in extinction probabilities whereas those in the second set are very similar.

The second feature is that life histories with a higher growth rate implied by their average projection matrix have higher finite-time extinction probabilities. As noted above in regard to ultimate extinction probabilities, this type of relationship is potentially of great applied significance.

Of more general importance is that the transient dynamics have implications that relate to common beliefs about the necessary structure of life history theories, about the time scale of ecological observations, and about the mechanisms of population regulation in most species.

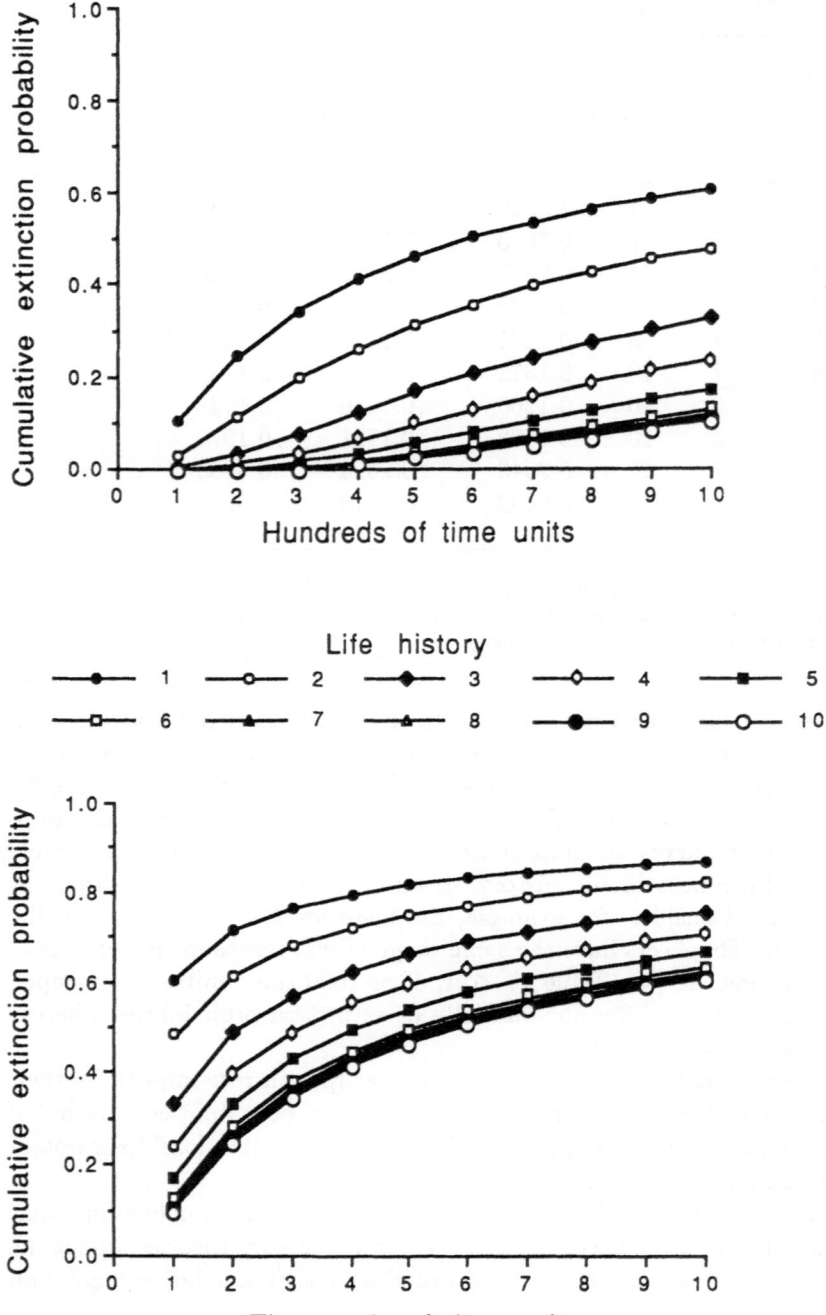

Figure 6. Transient extinction probabilities for the life histories in Table 1. $a = 0.0$ for all life histories and $x_0 = \ln N_0^*$ (see Table 3). These curves are independent of the value of ρ. Values are based upon Eq. (11) in Lande and Orzack (1988).

What is population stability?

The first implication concerns the belief that asymptotic stability of population number is an essential feature of models of life history evolution. (I use stability to mean the survival of the population as opposed to stability of numbers.) Of course, some life history theories do not rest upon this assumption (e.g., the theory of r - K selection, see MacArthur and Wilson 1967). Nonetheless, it is common to see statements or implications to the effect that density-independent models cannot be realistic because they predict only population extinction or explosion (e.g., see Klomp 1962 and Goodman 1984). (Of course, this criterion for realism "cuts both ways" in that some density-dependent models of population growth also predict inevitable extinction; see below.) There is an empirical reason and a theoretical reason why this argument is not dynamically meaningful.

The empirical reason relates to our present knowledge about population regulation. It is clear that there are populations in which density-dependence occurs and may even result in numerical stability and there are populations in which there is no evidence of density-dependent regulation of abundance (see Stiling 1988, Hassell et al. 1989, and Woiwod and Hanski 1992). The point is not to argue about the nature and quality of these data but rather to note that there is no compelling biological reason to *necessarily* regard either mode of population regulation as the most realistic "template" for life history evolution. The necessary inclusion of density-dependent population regulation in models of life history evolution would be appropriate if there were evidence that this process is *always* involved in determining the direction and outcome of life history evolution. There is no such evidence at present.

What is the theoretical reason? Suppose one *were* to regard population stability as an essential feature of models of life history evolution. The results described above reveal how this criterion is ambiguous. After all, populations that eventually must go extinct (since $a = 0.0$) may survive for very long periods of time. For example, the life histories in Figure 6 must go extinct, yet there is at least a 40% probability that a population containing any one of them is extant after 1000 time units. Indeed, it is straightforward to show that a population has $\approx$ 4% chance of being extant at a time equal to 1000 times the modal lifetime (see Table 6). So, for example, life history 10 has a modal lifetime of 880 time units. This means that a population has $\approx$ 4% chance of being extant after 880,000 time units! Even life history 1 has this same chance of surviving for 90,000 time units. This distinction between asymptotic and finite-time stability takes on ecological significance when one realizes that there are no time series of abundances for natural populations that come close to these lengths (even if the time unit is taken to be a month). To this extent, the survival of populations in *present* time series does not by itself indicate anything about the nature of population regulation. These calculations do not establish density-independent models as being of more relevance to natural populations than are density-dependent models. Instead, they simply show the potential relevance of such models (see also Den Boer 1991) and reveal the incorrectness of statements or implications to the effect that population stability over "several generations" is clear evidence for "some kind of negative feedback process"

Table 6. Probabilities that a population survives for times defined relative to the modal extinction time.

Time	Probability	Life History	Modal Extinction time
mode	0.916	1	90
2× mode	0.778	2	168
10× mode	0.418	3	321
50× mode	0.201	4	471
100× mode	0.097	5	622
1000× mode	0.044	6	772
		7	811
		8	840
		9	863
		10	880

The life histories are shown in Table 1. The probabilities are independent of the life history and of the initial population size. They are derived from Eq. (11) in Lande and Orzack (1988) and Eq. (9). Modal extinction times are for $a = 0.0$ and $x_0 = \ln N_0^*$ (see Table 3). Times (in arbitrary units) are rounded to the nearest integer.

(e.g., see p. 883 of Hassell et al. 1989). These calculations also clearly reinforce the importance of Connell and Sousa's (1983, p. 792) statement that "Appropriate scales of time and space must also be specified for the observations of the responses of the populations, before meaningful judgements concerning stability or persistence can be made."

A more general point in regard to the belief that population stability is a necessary feature of models of life history evolution is that many populations are known to be ephemeral in the "short-term" (e.g., see Schoener and Spiller 1987) and that all species are so in the "long-term." If anything, these data might cause one to regard population *instability* as essential to models of life history evolution. This, however, would be a mistake of the kind I have just criticized. These results concerning extinction dynamics simply serve to demonstrate that only data on the distribution of population longevities and on *mechanisms* of population regulation can serve to establish the greater relevance of either density-independent or density-dependent models to nature. In the absence of such data, statements stressing the exclusive necessity of density-independence or density-dependence are simply statements of faith.

Why "logic" does not tell us about population regulation in nature

The results I've described that concern asymptotic and finite-time dynamics all relate to populations that have a stochastic growth rate of 0.0. This value is taken to reflect an ecological balance – a balance between the environment and the population to the extent that the population has only the potential to maintain itself. (The word "potential" serves to highlight the distinction between this compatibility of the population with the environment and the fact that such populations must eventually cross an extinction boundary, see above). What are we to make of this focus on populations that have a stochastic growth rate of 0.0? One common claim is that such a focus is unrealistic because the attainment of such a growth rate is improbable. For example, Hassell et al. (1989) write that density-dependent regulation must be involved in the dynamics of "stable" populations "except in the *very unlikely* event that density-independent gain and loss happen to be in almost exact balance over the period" [emphasis added]. After all, the "probability" that a density-independent population possesses a growth rate of 0.000 ... is infinitesimally small. Since the one growth rate compatible with stability is improbable and all other stochastic growth rates lead to population extinction or explosion, this argument is said to provide a general reason why populations are almost certainly under density-dependent regulation. For example, Godfray and Hassell (1992, p. 673) apparently have this argument in mind when they write "It is a logical necessity that any population of plants or animals that persists in the environment must experience some form of density-dependent feedback ...." That even a stochastic growth rate of 0.0 results in extinction has been ignored (see below) as has been the fact that inevitable extinction is also a feature of some stochastic density-dependent models (see Tier and Hanson 1981). Even if one ignores these inconsistencies, this argument is completely fallacious and the reason can be seen by reconstructing its underlying logic.

The implied basis of this assessment of probability is a distribution of stochastic growth rates which either encompasses positive and negative values or more simply, just positive growth rates. The distribution is assumed to be broad enough that the probability of a growth rate being in the neighborhood near 0.0 is "very unlikely." (I refer to a neighborhood because *any* specific value of a continuous random variable has a zero probability of occurring. Of course, this is problematic in that the neighborhood must encompass nonzero growth rates, which lead to extinction or explosion). This argument is ultimately based upon the principle of indifference (cf., von Mises 1957) to the extent that the underlying broad (e.g., uniform) probability distribution is most likely thought to represent a *lack* of assumptions about the real world. Yet just as in the case of probabilistic inference in statistics, this use of the principle of indifference is wrong.

One reason is that such a principle should also be applied to a density-dependent model of population growth. The necessary conclusion is that it is "very unlikely" that a particular set of density-dependent checks on population growth averages out to be near 0.0. So, in this sense, density-dependent regulation of the population is also improbable. Of course, one might argue that what is required is that such an average only be broadly bounded. But again, even the probability that the average is within some bounds rests only upon a prior *arbitrary* decision as to the underlying probability distribution, e.g., that there is a uniform distribution of growth rates. The point here is that there is no free lunch. An arbitrary assumption as to the nature of underlying distribution of growth rates cannot substitute for the real thing – empirical information on the distribution.

There is another compelling biological reason why the "logical" argument demonstrating the necessity of density dependence is not meaningful. I've outlined how the principle of indifference is used to determine that a stochastic growth rate is 0.0 is improbable. This growth rate describes change in the *logarithm of population size*. There is no reason why this principle should not also be applied to changes in *population size*, i.e., to determine that a population growth rate of 1.0 is improbable. Yet these two equally well-motivated conditions can lead to a contradiction. This can be seen most simply by applying these results to a density-independent growth model of a scalar population,

$$N_{t+1} = \lambda_{t+1} N_t = \lambda_{t+1} \lambda_t \ldots \lambda_1 N_0 \tag{10}$$

where N_i denotes population size at time i and λ_t denotes the growth rate at time t. A mathematically equivalent representation is

$$\ln N_{t+1} = \sum_1^{t+1} \ln \lambda_t + \ln N_0 \tag{11}$$

Asymptotic stability of population size occurs when $E(\lambda) = 1.0$ or equivalently, when $E(\ln \lambda) = 0.0$, where the expectations are evaluated with respect to all of the environmental states. Of course, these expectations define only the behavior of the average of the process – even when they have a numerical value greater than zero, most populations may attain a very small size and go extinct (see Lewontin and Cohen 1969). This discrepancy between the average and typical behaviors highlights another ambiguity of the probabilistic argument concerning density-independence. No finite value of $E(\ln \lambda)$ or $E(\lambda)$ guarantees population persistence in a stochastic environment. (One could choose a value for, say, $E(\ln \lambda)$ that implies a lower probability of extinction than implied by a value of 0.0, but any choice must be based upon arbitrary assumptions as to what probability of extinction should be regarded as compatible with population stability. This is clearly oxymoronic.)

The stochastic growth rate for the population is

$$a = E(\ln \lambda) \quad t \to \infty \tag{12}$$

It follows from a Taylor series approximation to $E(\ln \lambda)$ (see above) that

$$a \approx \ln(E(\lambda)) - \frac{\sigma_\lambda^2}{2E(\lambda)^2} \tag{13}$$

where σ_λ^2 denotes the variance of λ. Yet, given $E(\ln \lambda) = a = 0.0$ and $E(\lambda) = 1.0$, the equality cannot be satisfied, since, of course, the variance is positive. The point is that the application of two equally well-motivated equalities leads to a contradiction. There is no biologically meaningful way in which the principle of indifference is more reasonably applied to growth rates on the arithmetic as opposed to the logarithmic scale (or vice-versa). There are ways to avoid the contradiction. For example, one might assume that additional terms of the Taylor series appear on the right hand side of Eq. (13). Perhaps these terms can satisfy the equality, i.e., their sum is equal to 0.0. Yet this way of avoiding the contradiction rests upon a prior arbitrary decision that the logarithmic growth rate is canonical. The point is not that a more complicated model of growth rate might resolve the contradiction but that it arises at all when the principle of indifference is applied to a population growth model. That this occurs indicates that the principle of indifference is an inappropriate basis for providing insights into nature.

Lest the reader feel that the principle of indifference deserves credence simply because of common use in population and evolutionary biology, I note that its problematic nature has been recognized elsewhere. For this reason, in phylogenetic inference the use of equal (i.e., uniform) transition probabilities among characters is not taken to represent a *lack* of assumptions (see Sober 1988, p. 228). The same assessment can be made about the use of equal weights for phenotypic characters. In population genetics, one could use the broad or uniform probability distribution argument used to demonstrate that density-independence is improbable to demonstrate that it is "very unlikely" that the selective value of a genotype is zero. Yet, this is not taken as a credible argument for a lack of selective neutrality. Finally, probabilities of events based upon analysis or simulation involving uniform-sampling procedures are not taken to describe nature. For example, Lewontin et al. (1978) determined via simulation that the fraction of randomly-constructed fitness matrices that result in a stable multiallelic polymorphism declines as the inverse square or cube of the number of alleles. Yet they correctly note (pp. 152–153) that this result is not to be taken to describe the *probability* that stable equilibria occur in nature or whether a particular multiallelic polymorphism is maintained by natural selection. There may be a very restricted set of fitness arrays in nature because those not resulting in stable polymorphism will tend not to be associated with polymorphism. (Indeed, Ginzburg (1979) showed that the fitness matrices that result in stable polymorphism are not a random sample but instead, have the feature that average heterozygote fitness exceeds average homozygote fitness.) So, just as the principle of indifference cannot be meaningfully used in these other biological contexts to motivate inferences about nature, it cannot be meaningfully used in the present context to demonstrate that most or almost all populations "must be" subject to density-dependence.

Population regulation: the necessity of data

What *can* we say about the distribution of population growth rates in nature? What little one can say suggests that the distribution must be highly nonuniform. This is especially so if one adheres to an "equilibrium" view of nature. After all, one should not expect to see any populations with growth rates that are less than 0.0 because, of course, they will have disappeared. The consequence is that half of the assumed distribution of growth rates is eliminated. One might also eliminate populations with very large growth rates. Surely, such populations will overrun their environment and cause the ecosystem to crash. Either of these emendations might ultimately be shown to have some more definite ecological basis (cf., Den Boer 1987). The point is, however, that their arbitrary nature highlights the artificiality of a probabilistic framework based upon a lack of information.

None of these arguments are intended to demonstrate the actual as opposed to potential relevance of density-independent population growth models to nature. As mentioned above, only information on long-term dynamics of populations and on mechanisms of population regulation can do this. To this extent, it is important to describe how density-independent populations are expected to behave. This is important in that it is often claimed that "observed" (but usually unspecified) time series of abundances are incompatible with a "random-walk," which is taken to be the prediction of a stochastic density-independent growth model (see above). (Of course, there are real populations whose fluctuations either match or *surpass* those predicted by such a model, see Den Boer 1990, 1991.)

Population boundedness

There is an apparent discrepancy between population "boundedness" and the prediction of the stochastic density-independent model discussed here in that the population must eventually reach every population size (including infinity) since the Wiener process is recurrent (see Bhattacharya and Waymire 1990). Yet, as with extinction dynamics, there are important ecological implications of the distinction between this asymptotic behavior and the finite-time behavior of the process.

Consider, for example, the life histories in Table 1 and assume, as in the analysis of extinction dynamics, that each characterizes a separate population with a stochastic growth rate of 0.0. The asymptotic lognormality of population size implies that the logarithm of the average population size at time t, $\ln(E(N_t))$, is approximately $x_0 + (a + \frac{1}{2}\sigma^2)t$ (Tuljapurkar and Orzack 1980). Since $\ln(E(N_t))$ changes in a clock-like manner and is *necessarily* increasing (even when $a = 0.0$), it exemplifies the potential for boundlessness of density-independent population growth, which is taken to be incompatible with observed abundances. This is, however, only a potential, as opposed to necessary, incompatibility for two reasons. First, the lognormality of population size implies that $\ln(E(N_t))$ is unrepresentative of the typical population size. Indeed, when $a = 0.0$, the logarithm of the modal population size is $x_0 - \sigma^2 t$, implying that the most probable result of density-independent fluctuations is population decline regardless of the life history (Tuljapurkar and Orzack 1980).

The second reason relates to population boundedness. Consider the probability that a population remains extant and does *not* attain a given size N^* larger than x_0. Up to a given time t, there are four possible outcomes for a given realization of the stochastic process:

1. The population goes extinct after not attaining N^*.
2. The population goes extinct after attaining N^*.
3. The population remains extant after not attaining N^*.
4. The population remains extant after attaining N^*.

Outcome 3 is the event of interest. Since this set of events is exhaustive, one has

$$P(3) = 1.0 - (P(1) + P(2) + P(4)) \tag{14}$$

where $P(x)$ denotes the probability of outcome x, the dependency on time t being implicit. But $P(1) + P(2)$ is equal to the total probability of extinction in time t and can therefore be estimated with the extinction time results described above. $P(4)$ is a conditional probability in that each sample path in this category must attain N^*

Table 7. Observed and expected probabilities that a population composed of individuals with life history 6 is extant and has not attained a population size in the time it takes $E(N_t)$ to attain the population size.

Population Size	Time	Probabilities Observed	Expected
20	56	0.191	0.178
100	945	0.547	0.527
1000	2218	0.543	0.536
5000	3108	0.512	0.518
10000	3491	0.503	0.509

Observed probabilities are estimates derived from stochastic simulation. For each probability, 10000 sample paths were generated independently. For each sample path, the analytical prediction for a is 0.0, $C \approx 1.043$, $\rho = 0.0$, and $x_0 = \ln N_0^*$ (see Table 3). Times (in arbitrary units) are rounded to the nearest integer. Any particular observed probability, P, has a standard deviation equal to $0.01\sqrt{[P(1.0 - P)]}$, which is always less than 0.005. Expected probabilities are derived from Eq. (15).

and then not attain the extinction boundary. This probability can be approximated by the unconditional probability that the process attains the upper boundary in time t. Accordingly,

$$P(3) \approx 1.0 - P(\text{extinction}|t) - P(\text{attainment of } N^*|t)) \qquad (15)$$

The last probability in Eq. (15) can be easily estimated since the first passage time dynamics that describe the extinction time dynamics of the Wiener process also describe the attainment of any value of population size given an initial logarithmic size x_0. (For example, in order to estimate the mean and variance of time to attainment of any N^* larger than x_0, one need only replace x_0 in Eq. (8) with $\ln N^* - x_0$.)

This approximation to $P(3)$ can be accurate to within a few percent as can be seen in Table 7 where estimates of $P(3)$ for life history 6 are shown along with probabilities of outcome 3 derived from stochastic simulation. The reason is that for short finite times the stochastic process only rarely attains a large upper boundary after attaining the lower boundary. Therefore, implicit inclusion of such sample paths by use of the unconditional probability instead of $P(4)$ usually results in only a small underestimate of $P(3)$. At longer times, this approximation to $P(3)$ can be as accurate as expected given its dependency on the dynamics of the Wiener process (see above and Lande and Orzack 1988). The only exceptions are some combinations of life histories (e.g., life history 10 in Table 1) and values of x_0 and $\ln N^*$ such that the stochastic dynamics of the Wiener process do not apply by the time sample paths encounter the boundaries. In these instances, Eq. (15) can be a large ($\geq 30\%$) overestimate of $P(3)$ as compared

Table 8. Probabilities that a population is extant and has not attained a given size in the time it takes $E(N_t)$ to attain the size.

Life History		Population Size				
		20	100	1000	5000	10000
1		0.393	0.433	0.423	0.408	0.401
	time =	70	231	463	625	694
2		0.344	0.487	0.480	0.463	0.454
		61	304	651	893	998
3		0.264	0.513	0.514	0.496	0.487
		57	461	1039	1444	1618
4		0.222	0.522	0.526	0.508	0.499
		56	621	1431	1996	2240
5		0.196	0.525	0.533	0.515	0.505
		56	783	1824	2551	2864
6		0.178	0.527	0.537	0.518	0.509
		56	945	2218	3108	3491
7		0.286	0.508	0.506	0.488	0.480
		175	1228	2734	3787	4240
8		0.345	0.487	0.479	0.462	0.454
		310	1525	3262	4477	5000
9		0.365	0.403	0.369	0.341	0.336
		459	1836	3806	5184	5777
10		0.288	0.241	0.187	0.164	0.148
		618	2158	4360	5899	6562

$x_0 = \ln N_0^*$ (see Table 3). Times (in arbitrary units) are rounded to the nearest integer. Probabilities are derived from Eq. (15) except those for life histories 9 and 10 which are derived from stochastic simulation (see Table 7 for general details).

to probabilities derived from simulations (especially for larger values of t) because predicted finite-time extinction probabilities ($P(1) + P(2)$) for such life histories are too small (Orzack, unpublished). The virtue of Eq. (15) is that estimates of $P(3)$ are easy to calculate for the same reason that finite-time extinction probabilities are easy to calculate (see above). (This "occupation time" problem for the Wiener process can be solved in a way that leads to less simple expressions, see Feller 1971, pp. 477–478.)

How often then do populations remain bounded between an upper population size and a lower extinction boundary in a given finite time? Table 8 shows the probability of Outcome 3 for upper population sizes of 20, 100, 1000, 5000, and 10000. The time interval considered for each size is the time it takes $E(N_t)$ to attain that size given a, σ^2, and x_0 for the life history in question. So, for example, for life history 4, 2240 time units elapse before $E(N_t)$ attains the value 10000. There is approximately a fifty percent probability that a given population is extant but has not attained the upper boundary by this time.

These results show that there is no *inevitability* of explosion in finite time for a population experiencing density-independent environmental fluctuations. Again, they do not serve to establish stochastic density-independent models as more than *potentially* relevant to real populations. The explanatory role of such models depends upon the demonstration that they more accurately describe the dynamics of real populations than do density-dependent models. In this regard, it is important to note that some features of the extinction dynamics associated with each type of model can be quite distinct. For example, Tier and Hanson (1981) analyze some stochastic density-dependent models of scalar populations and show that the mean and variance of extinction time are independent of population size as long as the population is not too small (compare with Eq. (8)). Presumably, this type of independence will hold true for nonscalar populations and thereby will represent a distinct difference between the two model types that can facilitate choice of one or the other as more relevant to nature (see Lande and Orzack 1988 and Dennis et al. 1991 for : .ated comments). In any case, the results concerning population boundedness and extinction serve to show that there is nothing about the dynamics of density-independence that is *inherently* incompatible with our present understanding of the real world.

The previous analyses concerning the ecological consequences of dispersed reproduction have a common demographic context: age structure. The analyses indicate that dispersed reproduction is not inherently advantageous in a variable environment. To this extent, there is no single evolutionary response to environmental variability that can be identified theoretically. Rather than being a cause for concern about the demise of an ecological generalization, this conclusion should be taken to reflect how much more sophisticated our understanding of population dynamics in variable environments has become. With the analytical framework outlined above one can gain an unparalleled view of some of the within- and between-population consequences of changes in both the structure and variability of life histories. This ability reflects the power of understanding dynamics. In the final section of this paper, I explore a different area in which the additional power of dynamical understanding is revealed.

Comparative demography

As mentioned above, there is a universal tension between present and future reproduction. On the one hand, present reproduction affords an advantage in that offspring can themselves reproduce sooner. On the other hand, future reproduction affords an advantage in that offspring may fare better with an older parent or may gain by avoiding environmental variability. Seen in this way, life history phenotypes such as iteroparity, diapause, seed banks, and prereproductive delays have much in common. So, for example, the delayed reproduction of an iteroparous life history represents a banking of offspring in much the same way that a seedbank does.

Life History	Phenotype	Projection Matrix	Type(s) of individuals in life history class	
			N_1	N_2
1	iteroparous $\alpha = 1,\ \omega = 2$ $f \equiv$ age 1 reproductive fraction	$\begin{bmatrix} fm_1 & (1-f)m_2 \\ s_1 & 0 \end{bmatrix}$	age 1 reproductive	age 2 reproductive
2	biennial $f \equiv$ flowering fraction	$\begin{bmatrix} 0 & fm_2 \\ s_1 & (1-f)s_2 \end{bmatrix}$	age 1 non-reproductive	age $2,\ldots,\ldots$ reproductive and non-reproductive
3	prereproductive delay $f \equiv$ reproductive fraction	$\begin{bmatrix} (1-f) & m_2 \\ fs_1 & 0 \end{bmatrix}$	age $1,2,\ldots,\ldots$ non-reproductive	age $2,\ldots,\ldots$ reproductive
4	diapause $f \equiv$ direct-developing fraction	$\begin{bmatrix} 0 & (1-f) \\ m_1 s_1 & fm_2 \end{bmatrix}$	age 1 reproductive	age 1,2 reproductive and non-reproductive

Figure 7. The structure of life histories with different kinds of reproductive delay. For a given life history, f is the average value of the reproductive delay. m_1 and m_2 are the fertilities of class 1 and class 2 individuals, respectively. s_1 and s_2 are the survivorships associated with the transitions between class 1 and class 2 and between class 2 and class 2, respectively.

The analytical framework outlined above allows one to determine the dynamical similarities and differences among such life histories. What causes the differences? Consider the matrices shown in Figure 7, which describe the transitions between life history classes. These classes are general relative to the age classes associated with a "standard" iteroparous life history in that, for example, the N_1 class for the prereproductive life history contains individuals of all ages. The same sort of temporal heterogeneity occurs in the N_2 class for this life history. As a result, such a life history has a long-term "memory" that is absent from the two-age class iteroparous life history. Note that all of these life histories satisfy the ergodicity condition that assures the applicability of Eq. (4) because each has two "reproductive" events even though for some only one class produces true newborns.

Important aspects of the dynamics associated with these life histories have been presented by Roerdink (1988, 1989), Tuljapurkar (1990), and Tuljapurkar and Istock (in press). For all of these life histories, one can show that some constant degree of reproductive delay (banking) is advantageous in an environment in which the reproductive rate (e.g., m_1 or m_2 in Figure 7) varies (see, e.g., Roerdink 1988). I consider here two different questions: what are the dynamical consequences of variation in the delay fraction? How do these consequences differ among the life histories in Figure 7?

The first step in answering these questions is to construct approximations to the stochastic growth rate a. Using Eq. (4), they are:

1. iteroparous:

$$a \approx \ln \lambda_0 - \frac{\sigma_f^2}{2} \left[\left(\frac{\partial \ln \lambda_0}{\partial f m_1} \right) m_1 - \left(\frac{\partial \ln \lambda_0}{\partial (1-f) m_2} \right) m_2 \right]^2 \qquad (16)$$

2. biennial:

$$a \approx \ln \lambda_0 - \frac{\sigma_f^2}{2} \left[\left(\frac{\partial \ln \lambda_0}{\partial f m_2} \right) m_2 - \left(\frac{\partial \ln \lambda_0}{\partial (1-f) s_2} \right) s_2 \right]^2 \qquad (17)$$

3. prereproductive delay:

$$a \approx \ln \lambda_0 - \frac{\sigma_f^2}{2} \left[\left(\frac{\partial \ln \lambda_0}{\partial (1-f)} \right) - \left(\frac{\partial \ln \lambda_0}{\partial f s_1} \right) s_1 \right]^2 \qquad (18)$$

and

4. diapause:

$$a \approx \ln \lambda_0 - \frac{\sigma_f^2}{2} \left[\left(\frac{\partial \ln \lambda_0}{\partial (1-f)} \right) - \left(\frac{\partial \ln \lambda_0}{\partial f m_2} \right) m_2 \right]^2 \qquad (19)$$

where f and σ_f^2 are the mean and variance of the delay fraction and λ_0 is the dominant eigenvalue of the average projection matrix. s_1, s_2, m_1 and m_2 are constant and are defined with respect to each matrix in Figure 7. These equations are qualitatively correct when $0 < f < 1$ and are expected to be quantitatively accurate when $0.1 \le f \le 0.9$.

For numerical analysis, values for the matrix entries have been chosen so as to maintain a constant value of λ_0 across life histories (see Table 9). Accordingly, the following analyses relate to the differences among life histories in their responses to environmental variability, there being no differences in their average growth rates.

As in the analysis of dispersed reproduction in age-structured life histories, one can construct indifference curves for the four life histories (see Figure 8). The indifference curves reveal significant differences among these life histories. So, for example, the life histories differ in whether an intermediate ($f = 0.5$) or extreme ($f = 0.1$ or 0.9) delay fraction is associated with the highest coefficient of variation for f. As a result, all other things being equal, one would predict differing directions of selection on reproductive delay in populations characterized by these different life histories.

The overall relationships between stochastic growth rate and coefficient of variation also differ among the life histories (see Figure 8). So, for example, life histories 2 and 4 differ in how the value of f affects their sensitivity to variability in f. For life history 2, a change from $f = 0.1$ to $f = 0.9$ results in an approximately ten-fold decrease in the

Table 9. Life histories with four types of reproductive delay (see Figure 7).

Life History	Phenotype	f	m_1	m_2	s_1	s_2	N_0^*
1	iteroparous	0.1	1.220	1.220	0.8	0.0	9.584
		0.5	1.111	1.111	0.8	0.0	12.461
		0.9	1.021	1.021	0.8	0.0	16.641
2	biennial	0.1	0.0	3.502	0.8	0.8	8.437
		0.5	0.0	1.500	0.8	0.8	8.750
		0.9	0.0	1.278	0.8	0.8	8.958
3	pre-reproductive delay	0.1	–	1.251	0.8	0.0	9.817
		0.5	–	1.250	0.8	0.0	9.333
		0.9	–	1.250	0.8	0.0	9.052
4	diapause	0.1	1.220	1.220	0.8	–	9.871
		0.5	1.111	1.111	0.8	–	9.231
		0.9	1.021	1.021	0.8	–	8.302

All sets of values share a constraint on the growth rate associated with the average projection matrix ($\lambda_0 = 1.0001$). – indicates that the relevant parameter is not defined for the matrix (see Figure 7). N_0^* is adjusted based upon the assumption that 10 class 1 individuals initiate a population. Values of m_1, m_2, and N_0^* are rounded.

range of C values associated with the change from $a = \ln \lambda_0$ to $a = 0.0$. For life history 4, the same change in f results in an approximately two-fold decrease in this range.

The effects of the different kinds of reproductive delay on transient extinction dynamics are shown in Figure 9. Life history 1 has the lowest probability of extinction while life history 2 has the highest. The ordering of extinction probabilities is temporally invariant (except, of course, when all populations have gone extinct). Nonetheless, the extinction probabilities at a given time are always within a few percent of one another. These similarities occur despite the significant differences in the structure of the life histories (cf., Figure 7).

This exercise in comparative demography is heuristic in that it involves comparison of a range of life history phenotypes that transcends the range found in any single species. Nonetheless, I have presented it to reveal how the dynamical framework outlined here allows one to compare even highly distinct life histories in a relatively easy manner.

Overview

All of the analyses presented here are partial with respect to their coverage of particular biological phenomena as well as with respect to interpretation. Accordingly, it is

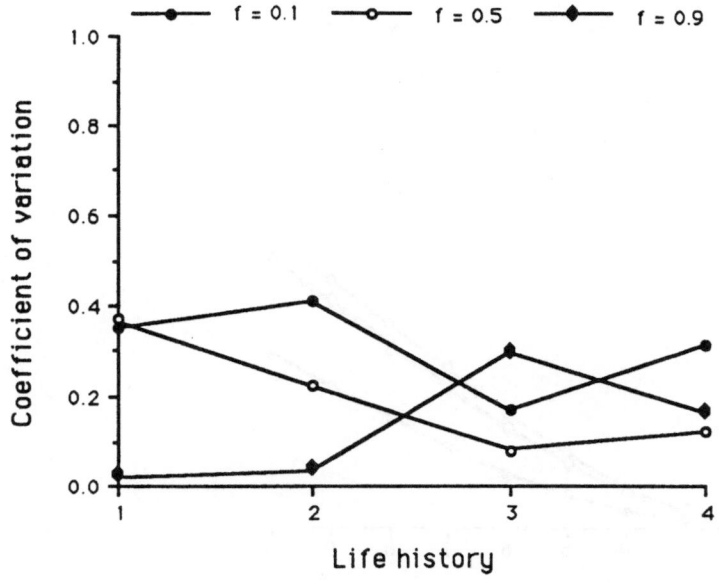

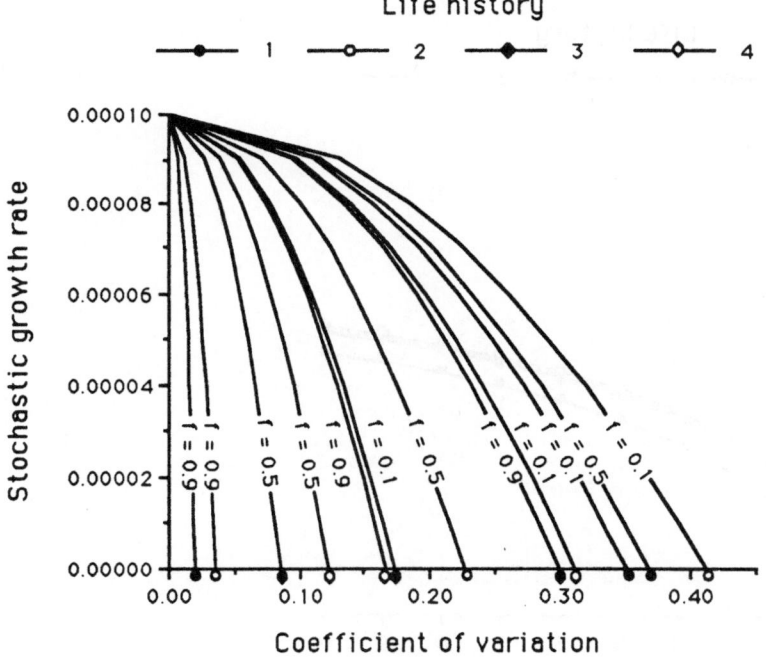

Figure 8. (top) Indifference curves for the life histories in Table 9 for a range of values of the reproductive delay fraction (f). $a = 0.0$ for all curves. Values of the coefficient of variation of f are derived from Eqs. (16), (17), (18), and (19). (bottom) The relationship between coefficient of variation and stochastic growth rate for the life histories in Table 9 for a range of values of f. The curves are derived from Eqs. (16), (17), (18), and (19).

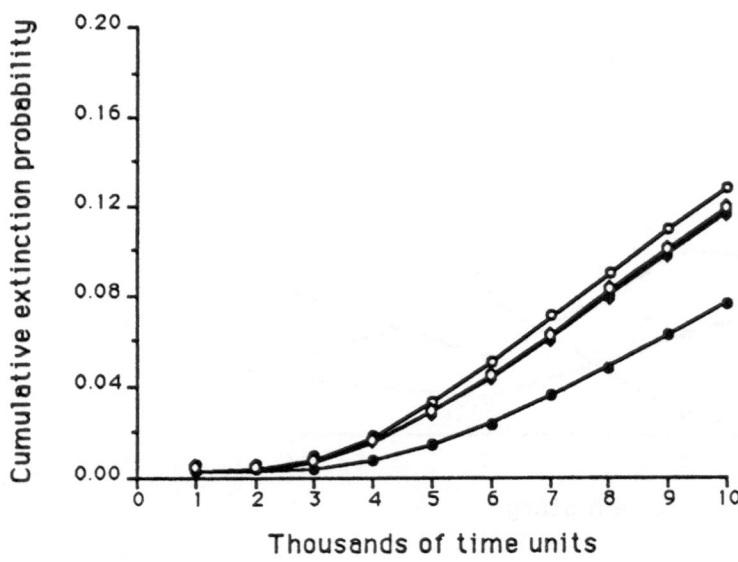

Thousands of time units

Life history

—•— 1   —○— 2   —◆— 3   —◇— 4

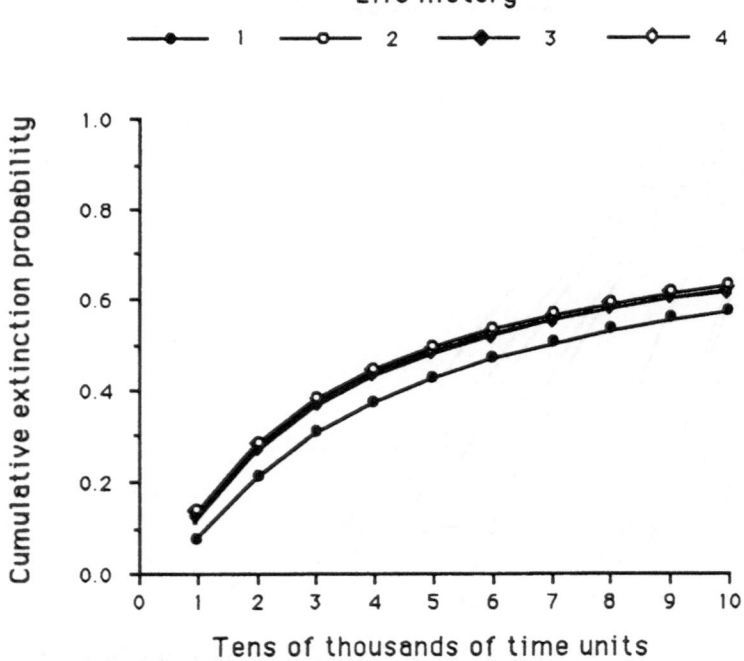

Tens of thousands of time units

Figure 9. Transient extinction probabilities for the life histories in Table 9 ($f = 0.5$). $a = 0.0$ for all life histories and $x_0 = \ln N_0^*$ (see Table 9). Values are based upon Eq. (11) in Lande and Orzack (1988).

clear that there is much work to be done simply in regard to exploration of population dynamics in variable environments. The analytical and interpretive methods outlined in this paper constitute a powerful set of tools that can be used in this effort. More generally, if both heuristic and empirical application of this framework continues, evolutionary biologists can more reasonably hope to fully understand life history evolution in variable environments. In this sense, the field of stochastic demography is at an exciting stage in its development and it is my hope that this chapter communicates this excitement.

Acknowledgements

I would like to thank Patrick Billingsley, Hal Caswell, Brian Charlesworth, Joel Cohen, Michael Neumann, Elliott Sober, and Jin Yoshimura for advice or information, Shripad Tuljapurkar for permission to cite unpublished work, two reviewers, Joe Bernardo and Mark Gustafson, for providing useful criticisms, and Colin Clark and Jin Yoshimura for inviting me to speak at the workshop on stochastic environments at the 1991 ESEB meeting in Debrecen. Work on this paper was carried out while the author was a visiting scientist in the Department of Entomology, Ohio State University and I would like to thank Dana Wrensch and Jerry Downhower for hospitality and space during this visit. I would also like to thank Brian Charlesworth, Marty Kreitman, NIH, and the Department of Ecology and Evolution at the University of Chicago for support.

Note

All of the computer programs used for the numerical calculations in this paper are available from the author.

References

Belovsky, G. E. 1987. Extinction models and mammalian persistence. In *Viable populations for conservation* (M.S. Soule (ed.), pp. 35–57. Cambridge, Cambridge University Press.

Bhattacharya, R. N., and E.C. Waymire. 1990. *Stochastic Processes with Applications*. New York, John Wiley & Sons.

Bierzychudek, P. 1982. The demography of jack-in-the pulpit, a forest perennial that changes sex, *Ecol. Monog.* 52: 335–351.

Caswell, H. 1978. A general formula for the sensitivity of population growth rate to changes in life history parameters. *Theor. Pop. Biol*, 14: 215–230.

Caswell, H. 1989. *Matrix population models*. Sunderland, MA, Sinauer Associates, Inc.

Charlesworth, B. 1980. *Evolution in age-structured populations*. Cambridge, Cambridge University Press.

Chhikara, R. S., and J.L. Folks. 1989. *The Inverse Gaussian Distribution: Theory, Methodology, and Applications*. New York, Marcel Dekker.

Cohen, J. 1977. Ergodicity of age structure in populations with Markovian vital rates, III. Finite-state moments and growth rate; an illustration, *Adv. App. Prob.* 9: 462–475.

Cohen, J. 1979a. Long-run growth rates of discrete multiplicative processes in Markovian environments, *J. Math. Anal. Appl.* 69: 243–257.

Cohen, J. 1979b. Comparative statics and stochastic dynamics of age-structured populations, *Theor. Popul. Biol.* 16: 159–171.

Cole, L. C. 1954. The population consequences of life history phenomena, *Quart. Rev. Biol.* 19: 103–137.

Connell, J. H., and W.P. Sousa. 1983. On the evidence needed to judge ecological stability or persistence, *Amer. Nat.* 121: 789–824.

Cox, D. R., and H.D. Miller. 1965. *The theory of stochastic processes*. London, Chapman and Hall.

Den Boer, P. J. 1987. Density dependence and the stabilization of animal numbers 2. The Pine Looper, *Neth. J. Zool.* 37: 220–237.

Den Boer, P. J. 1990. Density limits and survival of local populations in 64 carabid species with different powers of dispersal, *J. Evol. Biol.* 3: 19–48.

Den Boer, P. J. 1991. Seeing the trees for the wood: random walks or bounded fluctuations of population size? *Oecologia* 86: 484–491.

Dennis, B., P.L. Mulholland, and J.M. Scott 1991. Estimation of growth and extinction parameters for endangered species, *Ecol. Monog.* 61: 115–143.

Diamond, J. M. 1984. "Normal" extinctions of isolated populations. In: *Extinctions* (M. Nitecki (ed.), pp. 191–246. Chicago, University of Chicago Press, Chicago.

Emlen, J. M. 1988. Evolutionary ecology and the optimality assumption. In *The Latest on the Best* (J. Dupre (ed.), pp. 163–177. Cambridge, MIT Press.

Feller, W. 1971. *An Introduction to Probability Theory, II*. New York, John Wiley & Sons.

Furstenberg, H., and H. Kesten. 1960. Products of random matrices, *Ann. Math. Stat.* 31: 457–469.

Ginzburg, L. R. 1979. Why are heterozygotes often superior in fitness? *Theor. Pop. Biol.* 15: 264–267.

Godfray, H. C. J., and M. P. Hassell. 1992. Long time series reveal density dependence, *Nature* 359: 673–674.

Goodman, D. 1984. Risk spreading as an adaptive strategy in iteroparous life histories, *Theor. Pop. Biol.* 25: 1–20.

Hassell, M. P., J. Latto, and R. M. May. 1989. Seeing the wood for the trees: detecting density dependence from existing life-table studies, *J. Anim. Ecol.* 58: 883–892.

Hines, W. G. S. 1980. Strategy stability in complex populations, *J. Appl. Prob.* 17: 600–610.

Hines, W. G. S. 1982. Mutations, perturbations and evolutionarily stable strategies, *J. Appl. Prob.* 19: 204–209.

Hines, W. G. S. 1990. A discussion of evolutionarily stable strategies. In *Mathematical and Statistical Developments of Evolutionary Theory*, S. Lessard (ed.), pp. 229–267, Norwell, MA, Kluwer Academic Publishers.

Holgate, P. 1967. Population survival and life history phenomena, *J. Theor. Biol.* 14: 1–10.

Jinks, J. L., and H S. Pooni. 1988. The genetic basis of environmental sensitivity. In *Proceedings of the Second International Conference on Quantitative Genetics*, B. S. Weir, M. M. Goodman, E. J. Eisen, and G. Namkoong (eds.), pp. 505–522, Sunderland, MA, Sinauer Associates, Inc.

Johnson, N. L., and S. Kotz. 1970. *Distributions in Statistics. Continuous Univariate Distributions – 1*. New York, John Wiley & Sons

Kendall, M., and A. Stuart. 1977. *The Advanced Theory of Statistics, 1. Distribution Theory*, New York, Macmillan.

Klomp, H. 1962. The influence of climate and weather on the mean density level, the fluctuations, and the regulation of animal populations, *Neth. J. Zool.* 15: 68–109.

Kolman, W. 1960. The mechanism of natural selection for the sex ratio, *Amer. Nat.* 94: 373–377.

Kuznetsov, P. I., R. L. Stratonovich, and V. I. Tikhonov. 1965. Correlation functions in the theory of the Brownian motion generalization of the Fokker-Planck equation. In *Nonlinear Transformations of Stochastic Processes*, P. I. Kuznetsov, R. L. Stratonovich, and V. I. Tikhonov (eds.), pp. 77–100, Oxford, Pergamon Press.

Lacey, E. P., L. Real, J. Antonovics, and D. G. Heckel. 1983. Variance models in the study of life histories, *Amer. Nat.* 122: 114–131.

Lande, R. 1987. Extinction thresholds in demographic models of territorial populations, *Amer. Nat.* 130: 624–635.

Lande, R. 1988. Demographic models of the northern spotted owl (*Strix occidentalis caurina*), *Oecologia* 75: 601–607.

Lande, R., and S. H. Orzack. 1988. Extinction dynamics of age-structured populations in a fluctuating environment, *Proc. Nat. Acad. Sci.* 85: 7418–7421.

Leslie, P. H. 1945. On the use of matrices in certain population mathematics, *Biometrika* 33: 213–245.

Leslie, P. H. 1966. The intrinsic rate of increase and the overlap of successive generations in a population of guillemots (*Uria aalge* Pont.), *J. Anim. Ecol.* 35: 291–301.

Levinton, J. S. and L. Ginzburg. 1984. Repeatability of taxon longevity in successive foraminifera radiations and a theory of random appearance and extinction, *Proc. Nat. Acad. Sci.* 81: 5478–5481.

Lewontin, R. C., and D. Cohen. 1969. On population growth in a randomly varying environment. *Proc. Nat. Acad. Sci.* 62: 1056–1060.

Lewontin, R. C., L. R. Ginzburg, and S. D. Tuljapurkar. 1978. Heterosis as an explanation of large amounts of genetic polymorphism. *Genetics* 88: 149–169.

MacArthur, R. H., and E. O. Wilson. 1967. *The Theory of Island Biogeography*. Princeton, Princeton University Press.

Murphy, G. 1968. Pattern in life history and the environment. *Amer. Nat.* 102: 391–403.

Orzack, S. H. 1985. Population dynamics in variable environments V. The genetics of homeostasis revisited, *Amer. Nat.* 125: 550–572.

Orzack, S.H., and S. Tuljapurkar. 1989. Population dynamics in variable environments VII. The demography and evolution of iteroparity, *Amer. Nat.* 133: 901–923.

Pollard, J. H. 1973. *Mathematical Models for the Growth of Human Populations*. Cambrdige, Cambridge University Press.

Roerdink, J. B. T. M. 1988. The biennial life strategy in a random environment, *J. Math. Biol.* 26: 199–215.

Roerdink, J. B. T. M. 1989. The biennial life strategy in a random environment. Supplement, *J. Math. Biol.* 27: 309–319.

Schaffer, W. M. 1974. Optimal reproductive effort in fluctuating environments, *Amer. Nat.* 108: 783–790.

Schoener, T. W., and D. A. Spiller. 1987. High population persistance in a system with high turnover, *Nature* 330: 474–477.

Sober, E. 1988. *Reconstructing the Past*. Cambridge, MA, MIT Press.

Stiling, P. 1988. Density-dependent processes and key factors in insect populations, *J. Anim. Ecol.* 57: 581–593.

Taylor, G. C. 1985. Primitivity of products of Leslie matrices, *Bull. Math. Biol.* 47: 23–34.

Thoday, J. M. 1953. Components of fitness, *Symp. Soc. Exp. Biol.* 7: 96–113.

Tier, C. and F. B. Hanson. 1981. Persistence in density dependent stochastic populations, *Math. Biosci.* 53: 89–117.

Tuljapurkar, S. D. 1982a. Population dynamics in variable environments II. Correlated environments, sensitivity analysis and dynamics, *Theor. Pop. Biol.* 21: 114–140.

Tuljapurkar, S. D. 1982b. Population dynamics in variable environments III. Evolutionary dynamics of r-selection, *Theor. Pop. Biol.* 21: 141–165.

Tuljapurkar, S. D. 1986 Demography in stochastic environments II. Growth and convergence rates, *J. Math. Biol.* 24: 569–581.

Tuljapurkar, S. D. 1989. An uncertain life: demography in random environments, *Theor. Pop. Biol.* 35: 227–294.

Tuljapurkar, S. D. 1990. Delayed reproduction and fitness in variable environments, *Proc. Nat. Acad. Sci.* 87: 1139–1143.

Tuljapurkar, S. D., and S. H. Orzack. 1980. Population dynamics in variable environments I. Long-run growth rates and extinction, *Theor. Pop. Biol.* 18: 314–342.

Tuljapurkar, S. D., and C. Istock. In press. Environmental uncertainty and variable diapause, *Theor. Pop. Biol.*

Vandermeer, J. H. 1975. On the construction of the population projection matrix for a population grouped in unequal stages, *Biometrics* 31: 239–242.

Vandermeer, J. H. 1978. Choosing category size in a stage projection matrix, *Oecologia* 32: 79–84.

von Mises, R. 1957. *Probability, Statistics, and Truth*. London, George Allen and Unwin.

Werner, P. A., and H. Caswell. 1977. Population growth rates and age versus stage-distribution models for teasel (*Dipsacus sylvestris* Huds.), *Ecology* 58: 1103–1111.

Whitmore, G. A. 1978. Discussion of the paper by Professor Folks and Dr. Chhikara, *J. Roy. Stat. Soc. Ser. B* 40: 285–286.

Woiwod, I. P., and I. Hanski. 1992. Patterns of density dependence in moths and aphids, *J. Anim. Ecol.* 61: 619–629.

PLASTICITY IN FLUCTUATING ENVIRONMENTS

Jesús Alberto León

Instituto de Zoología Tropical, Facultad de Ciencias
Universidad Central de Venezuela
Aptdo. 47058, Caracas 1041-A, Venezuela

1. Introduction

Most models of selection operating in fluctuating environments consider only rigid phenotypes. Genetic models (Haldane & Jayakar 1963, Gillespie 1973, Karlin & Liberman 1974) are also incomplete because they refer only indirectly to the phenotype. Strategic (optimality) analyses, on the other hand, usually look for an optimal fixed phenotypic compromise, presumably established by selection when each of the possible environmental states would favour a different type (Levins 1968a, Schaffer 1974, Oster & Wilson 1978, Real 1980, León 1983, Brown & Venable 1986, Yoshimura & Clark 1991). This optimum is obtained as follows. Any given phenotype X would exhibit an array of fitnesses $W_y(X)$ corresponding to different environmental states Y. The type endowed with maximal geometric mean M (or logarithmic expectation $E(\ln W)$) of the fitnesses, would eventually prevail (Gillespie 1973, León 1985, Frank & Slatkin 1990, Yoshimura & Clark 1991). This outcome is valid only in populations without age structure or density dependence since the presence of these factors would complicate the analysis (Caswell 1983, Bulmer 1984, 1985; Ellner 1985a, 1985b; Goodman 1984, Tuljapurkar 1982, 1989). The second order Taylor expansion of the logarithmic expectation gives (Gillespie 1977, León 1983, Real 1980) the approximation:

$$E(\ln W) = \ln E(W) - \frac{1}{2} \frac{V(W)}{E(W)^2},$$

where W is fitness, $E(W)$ its expectation and $V(W)$ its variance. Therefore, the constrained maximum of $E(\ln W)$ will be a compromise between maximizing the mean $E(W)$ and minimizing the variance $V(W)$. This would characterize the optimal phenotype.

This approach, of maximizing the geometric mean or the logarithmic expectations, has been used to tackle at least three classical problems: First, to ascertain whether changing coarse-grained environments will select for (rigid) generalists or (rigid) specialists (Levins 1968a). Second to ask what is the result of random change in effective fecundity or adult mortality on optimal (rigid) reproductive effort, as compared with reproductive effort in a constant environment (Schaffer 1974). Third, to understand the evolution of dormancy in uncertain environments (Cohen 1966). Variations on the latter theme are: the evolution of delayed reproduction (Cohen 1968), diapause (Cohen 1970, Hanski 1988) and dispersal (Venable & Lawlor 1980, Metz et al. 1983, Levin et al. 1984, Klinkhamer et al. 1987, Venable & Brown 1988). Typically, there is plasticity to adopt two extreme states (germinating vs. dormant seed, current vs. delayed reproduction ... etc.) but the probability of adopting either state is a rigid strategy in most models. The exception is a model by Cohen (1967) in which state-specific probabilities are allowed to change according to the presumed quality of each year as indicated by a signal.

Plasticity in changing environments, on the other hand, has rarely been considered in explicit mathematical formulations. For example, a review by Bradshaw (1965) was entirely verbal. Besides a few remarks by Levins (1963, 1968a,b) the few mathematical models available either treat particular problems in a detailed manner (e.g., Gross 1984, Ellner & Beuchat 1984, Kingsolver & Watt 1983) or consider some particular kind of plasticity (Cooper & Kaplan 1982, Kaplan & Cooper 1984).

Of course, several other models have been produced which refer to the evolution of plasticity in circumstances not reviewed here. Examples are: developmental plasticity in spatially varied environments (Lively 1986), changes of sex, including genetic vs. environmental sex determination (Charnov & Bull 1977, Bull 1981, 1983, 1987), facultative parthenogenesis (Lively 1987), and socially controlled plasticity in constant environments (Fagen 1987). Caswell (1983) also discusses some aspects of plasticity in age-structured models.

The focus of Stearns on reaction norms constitutes a recent advance in approaches to the kind of problems described above (Stearns & Koella 1986, Stearns 1989).

Here, I attempt to introduce phenotypic plasticity into the three classical themes mentioned at the outset. I do this by developing a unified approach provided by Levins's strategic analysis. Thus, I first put the three models in a Levins framework and then show how to bring plasticity into the picture. Since Levins's analysis is mostly graphical, I attempt to maintain the geometric mood throughout.

Besides the synthesis, which allows one to handle disparate problems using the same method, several aspects of section 4 (which is the core of the paper) are new: the distinction of ignorant versus informed plasticity, the explicit introduction of the cost of plasticity in fitness set analysis of fluctuating environments, and the use of "conditional" geometric means for studying imperfectly informed plasticity.

2. Models in a Levins framework

The main models mentioned above can be put in a Levins fitness set framework. I now show how this can be accomplished.

The Levins analysis (Levins 1968a) considers two environmental states, 1 and 2. Two bell-shaped curves, W_1 and W_2, assign fitnesses in the two environments to each value of a phenotypic variable X (Figs. 1A, 1B). In the usual representation, the two curves only differ in the location of their maxima (optimal phenotypes, $\widehat{X}_1$ and $\widehat{X}_2$) but not in their width or height. Differences in these, though, could be easily included (for instance: a high curve for a 'favorable' environment, and a low curve for an 'adverse' one).

A new graph is drawn, now using two fitnesses W_1 and W_2 as the coordinates of each phenotype X. The two curves are thus replaced by the frontier of a pear-shaped set called the fitness set (Figs 1a, 1b). The optimal phenotype in an environment changing independently each year between states 1 or 2 with probability q or p ($p + q = 1$), is easily obtained. It will be a point somewhere on the north-east frontier of the fitness set which touches tangentially that level curve ($M = $ const.) of the geometric mean M furthest apart from the origin of coordinates.

I shall call the north-east (or upper-right) frontier the "opportunity border". Such a border is concave downwards (i.e. seen from the origin of coordinates; Fig. 1a) when the two $W(X)$ curves are wide and/or their $\widehat{X}_1$ and $\widehat{X}_2$ do not differ too much (Fig 1A). If these differ enough and/or the curves are narrow (Fig. 1B) the border will be convex (Fig. 1b).

The geometric mean will be $M = W_1^q W_2^p$. The level curves $W_1^q W_2^p = $ const. will be equilateral hyperbolae if $p \approx q$. Otherwise they will tend to vertical lines as q approaches 1, or to horizontal lines as p approaches 1 (see Fig. 13.1 in Yodzis 1989). Thus, if the border is concave selection favors an intermediate optimum or 'rigid generalist' when both environmental states show up with similar frequency. If one environment predominates, the respective specialist is favored by selection.

Mixtures of the two specialists will be represented by the straight line going from one corner of the border to the other. If the frontier is concave, such a line lies below the border, but when the frontier is convex, the line lies above it hanging between the two horns of the fitness set. In this case, a mixture of specialists in the population would have a higher geometric mean (the mixture line touches tangentially a higher level curve) than any generalist if p is intermediate. Beginning with Levins (1968a), this has been interpreted as a polymorphism endowed with higher population fitness. Here, a different interpretation shall be preferred as will be argued in the next section.

Schaffer (1974) studied the effect of environmental variability on optimal reproductive effort. A reference environment exists in which the fitness of a perennial is $W_0 = F_0 + P_0$. Here, the effective fecundity $F_0(\epsilon)$ is an increasing function of the reproductive effort ϵ, whereas the adult interannual survivorship, $P_0(\epsilon)$, is a decreasing function. The optimal effort in this environment, $\hat{\epsilon}_0$, will be an intermediate value ($0 < \hat{\epsilon}_0 < 1$) when both $F_0(\epsilon)$ and $P_0(\epsilon)$ are concave. This gives a maximum of W_0 characterized by $(dW_0/d\epsilon) = 0$. If fluctuations in either F or P are introduced, the new optimal ϵ, say $\hat{\epsilon}$, corresponds to a maximum of the geometric mean M, and differs from $\hat{\epsilon}_0$. Schaffer (1974) considers symmetric environmental fluctuations switching between a good (g) or a bad (b) environment which occur with equal probability ($p = q$). Thus, if the random variation affects F, $W_g = F_0(1 + S) + P_0$ and $W_b = F_0(1 - S) + P_0$. This gives $\hat{\epsilon} < \hat{\epsilon}_0$, or a reduction of the optimal effort. On the other hand, if the variability impinges on P, $W_g = F_0 + P_0(1 + S)$ and $W_b = F_0 + P_0(1 - S)$. The maximum of the geometric mean is now located at $\hat{\epsilon} > \hat{\epsilon}_0$ and the optimal effort increases.

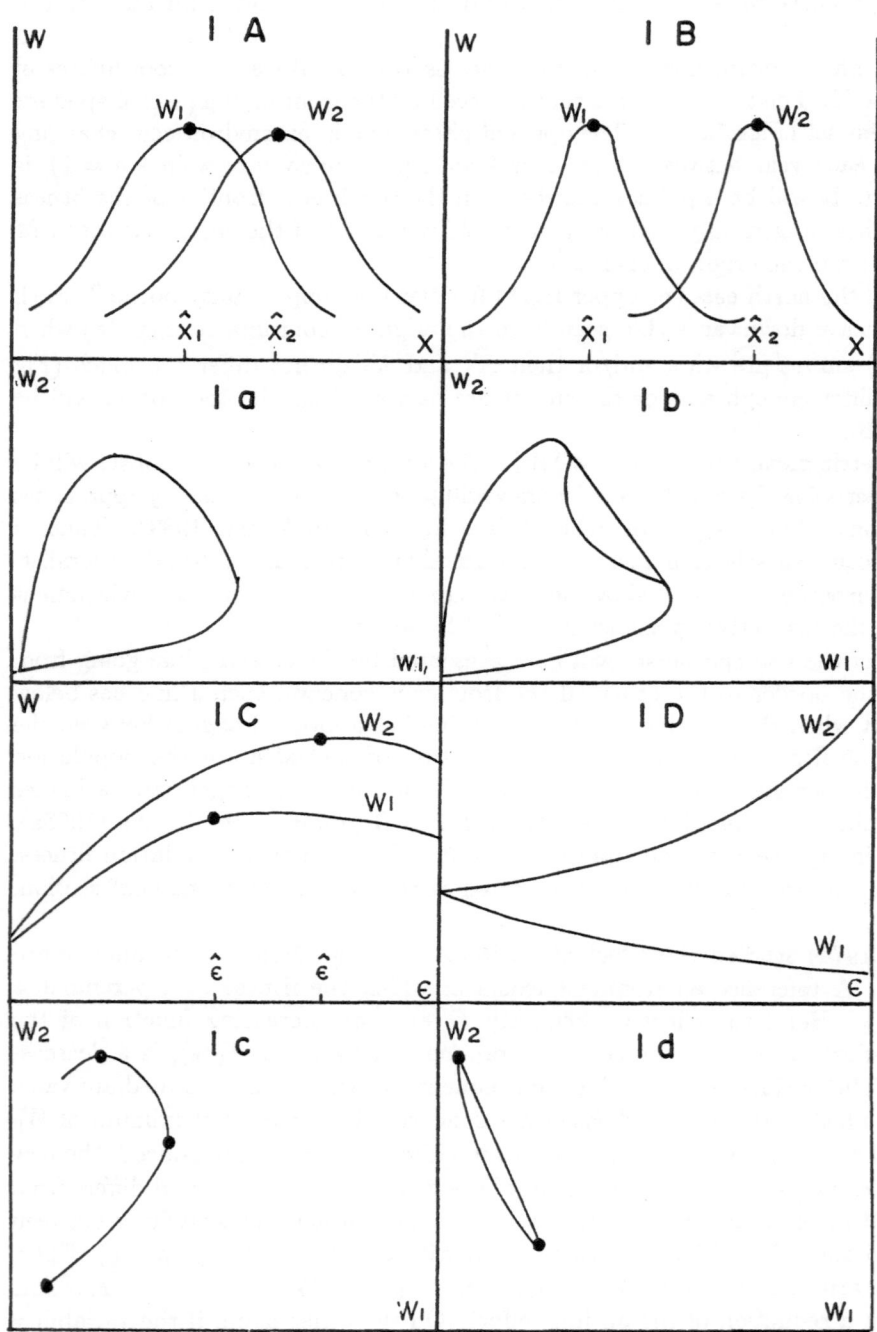

Figure 1

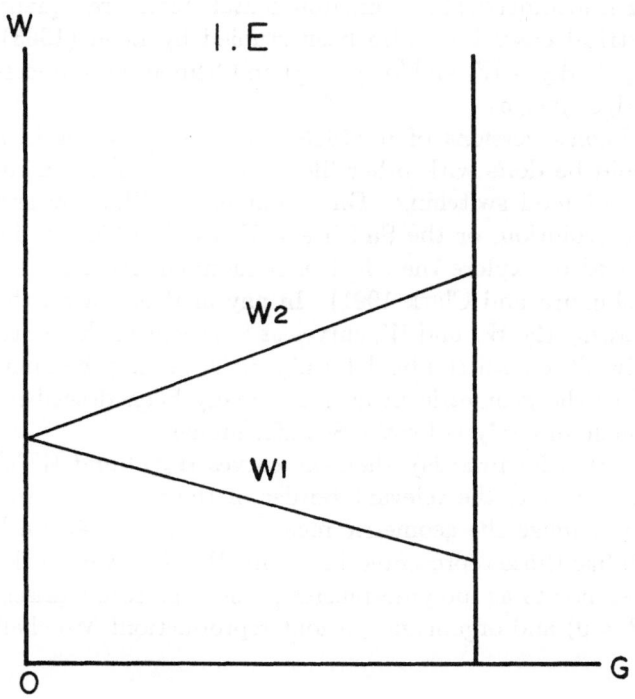

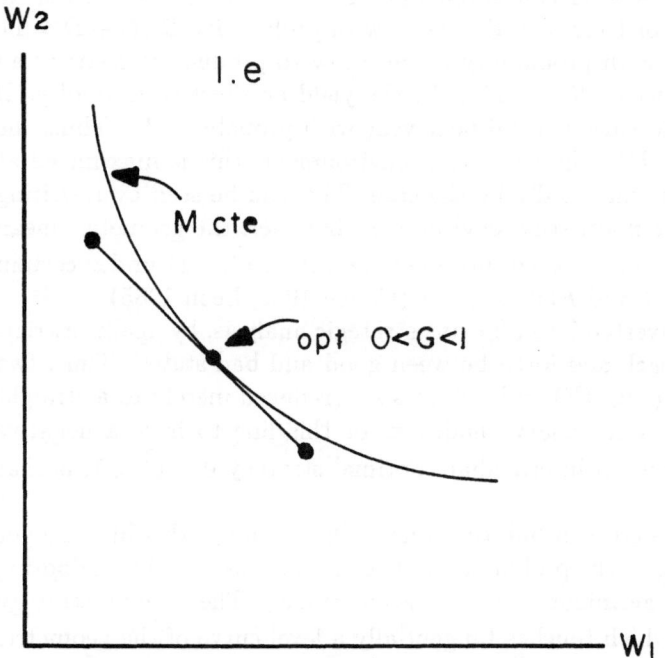

Figure 1

The situation considered by Schaffer (1974) has been branded 'neutral uncertainty' by León (1983) because switching is symmetrical and environmental states are equally probable $(p = q)$. Two asymmetrical cases have also been studied by León (1983): 'promising uncertainty' when $(W_g \gg W_0 > W_b$ and/or $p \gg q)$ and 'threatening uncertainty' when $(W_g > W_0 \gg W_b$ and/or $p \ll q)$.

Schaffer's model (1974), or León's versions of it (1983), can be easily set in a Levins framework. The same could be done with other life history models by introducing stochastic environmental bad-good switching. For example, the Charnov and Krebs (1973) model of clutch size evolution, or the Smith and Fretwell (1974) model of optimal offspring size, can be used to explore the effect of random change on these traits (McGinley et al. 1987, Yoshimura and Clark 1991). In any of these cases, the Levins analysis simply requires drawing the W_g and W_b curves as functions of the pertinent trait, with which the respective fitness-set can be determined. Obtaining the point of the set frontier which maximizes the geometric mean has already been described. Figures 1C, 1c and 1D, 1d show such an analysis for the Schaffer model.

If both $F(\epsilon)$ and $P(\epsilon)$ are convex (downwards), then the curves $W_g(\epsilon)$ and $W_b(\epsilon)$ will also be convex (Fig. 1D). As a result, the relevant border of the fitness-set will be convex and the candidates to maximize the geometric mean are 'mixed strategies': points of a 'mixture line' (Fig. 1d) like those represented in Figure 1b. However, in the conventional interpretation of these points as 'polymorphisms', a strange result arises: the mixture includes 'annuals' $(P = 0)$ and organisms without reproduction! We shall reconsider this view next.

Dan Cohen (1966) considered the evolution of seed dormancy as an adaptation to an uncertain environment. A seed can adopt one of two extreme behaviours: to germinate, with probability G, or to remain dormant, with probability D $(G + D = 1)$. If it germinates, it can survive with probability S and grow to produce B seeds as an adult annual plant. The product of S and B is F, the yield or effective fecundity. If the seed stays dormant, it could survive until next year with probability V. Thus, the fitness of a seed is $W = GF + DV$. In a constant environment, this is maximized at $G = 1$ as long as $F > V$, which will usually be the case. This can be seen by rewriting W as $W = V + (F - V)G$. In a fluctuating environment however, the geometric mean of the Ws is maximized. This may give an intermediate $\widehat{G}(0 < \widehat{G} < 1)$ under certain conditions, namely $E(F_t/V_t) > 1$ and $E(V_t/F_t) > 1$ (Cohen 1966, León 1985).

This model can also be converted into a Levins strategic analysis, by again restricting the environment to jump back and forth between good and bad states. Thus, two straight lines W_g and W_b exist (Fig. 1E) and a fitness-set (reduced merely to a straight line) can be drawn (Fig. 1e). A necessary condition for this line to have a negative slope and thus to possibly obtain an intermediate optimal strategy $0 < \widehat{G} < 1$, is that $F_b < V_b$.

Notice that the fitness-set here is a 'mixture line'. The points of this line are not 'polymorphisms'; instead they give the probabilities (G or D) of a seed flexibly adopting one of the extreme phenotypes (germinates or becomes dormant). The optimal strategy is, of course, that pair $(\widehat{G}, \widehat{D})$ which touches tangentially a level curve of the geometric mean. 'Promising uncertainty' ($p \rightarrow 1$, high F's) favors 'risk incurrence' ($\widehat{G} \rightarrow 1$),

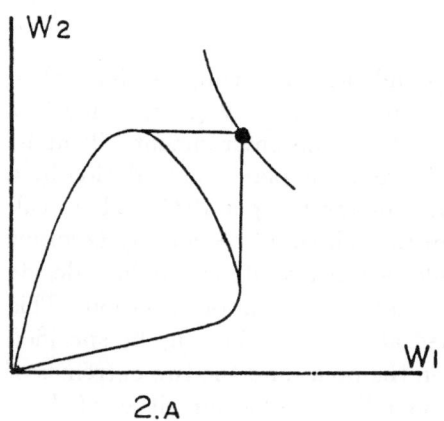

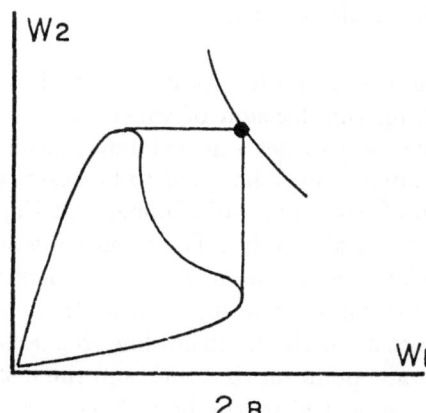

Figure 2. Panglossian plasticity.

while 'neutral' or 'threatening' uncertainty ($p \leq q$, low F's) favors 'risk avoidance' ($0 < \widehat{G} \ll 1$) (León 1983, 1985).

3. Panglossian plasticity

It has become fashionable in evolutionary theory to invoke Dr. Pangloss to refer to utopian scenarios. This character of Candide, the short novel by Voltaire, was intended as a caricature of 'public' Leibniz. Bertrand Russell (1945) distinguishes two Leibnizian philosophies: one, 'optimistic and shallow', published to please princesses; another, profound but hidden, left in his desk. It was the first Leibniz who proclaimed ours to be 'the best of all possible worlds'. I now mimic such a Panglossian view of plasticity.

Consider a perfectly plastic organism living in a changing environment. Suppose it recognizes unequivocally the environmental states as they appear. In this scenario, it will be optimal to change phenotype keeping the environmental pace (e.g., to adopt in each case 'the best of all possible' phenotypes according to the impinging circumstances). Thus, $\widehat{X}_1$ or $\widehat{X}_2$ will be exhibited whenever conditions 1 or 2 show up. In a fitness-set representation, such an organism is specified by a point of coordinates $(\widehat{W}_1, \widehat{W}_2)$. It escapes the constraint imposed by the set frontier (Fig. 2a), and it also escapes the 'mixture line', if the frontier is convex (Fig. 2b). Since this point is well beyond these limits, it can reach a much *higher* level curve of the *geometric mean* compared with any point contained within such limits.

Clearly, it is super-adaptive to be phenotypically flexible in a changing environment, because this strategy is superior to a rigid generalist, specialist or even a 'mixed strategy'. Why then, does this diversity of strategies exist? To answer, let me point out the absence in the Panglossian model of any resistance or difficulty which opposes plasticity such as the lack of costs and the too easy assumption of perfect information about the environment.

4. Realistic plasticity

Ignorant plasticity. Assume that plasticity is cheap and easy. But suppose that, when making the decision of which phenotype to adopt, there is no way (or it is hard or expensive) to get relevant information about the state of the environment. That is, the environment happens to be *inscrutable*. Then, it would be plausible to decide by a kind of 'coin flipping' (Cooper and Kaplan 1982, Kaplan and Cooper 1984). This I call *'ignorant plasticity'*. For example, a seed hidden in the soil could be entirely ignorant of what is going on in the environment which it would face as a seedling should it decide to germinate. The question is, then, what 'coin' to use in the random decision. This brings us to the 'mixture line' repeatedly mentioned above. A 'coin' may be specified by some point on that line, and the distance between the point and the two extremes of the segment gives the bias of the 'coin' (i.e. the probability of adopting either of those extreme phenotypes). Of course, this bias should be stipulated by natural selection. The optimal bias is obtained by finding the point on the 'mixture segment' which is tangent to a level curve of the geometric mean. This, instead of an 'optimal' polymorphism, is the interpretation given herein to optima lying in the 'mixture line'.

The 'mixture line' is external to the fitness-set only when the 'opportunity border' is convex (seen from the origin), or consists solely of the two points representing qualitatively different phenotypes lying at the extremes of the 'mixture line' (Figs. 1b, 1d, 1e). In all these cases, the optimal strategy is to exhibit *ignorant plasticity*, by choosing at random the individual phenotype for each time period and adopting one of the two extremes (one of the two optima $\widehat{X}_1$, $\widehat{X}_2$, in Levins's original model; Fig. 1b). Either to be an annual (with full reproduction and death, $\epsilon = 1$) or survive with maximal defense but without reproduction ($\epsilon = 0$) in the life-history model (Fig. 1d) or alternatively to either germinate or remain dormant in Cohen's model (Fig. 1e).

There is one case where the 'mixture line' resides within the fitness-set so that the set border reaches a higher geometric mean level curve. This happens, of course, if the border is concave. Therefore, even if plasticity is cheap and easy, *rigidity* is better when the random environment is *inscrutable*. This corresponds to a rigid generalist or specialist in Levins's concave model (Fig. 1a) and to a perennial life-history in Schaffer's concave case (Fig. 1c).

Costly plasticity. When the organism can obtain dependable cues about the changing environment, it faces a *decipherable environment*. Then it would be convenient to be flexible, to display *informed plasticity*. But plasticity can be costly, difficult to attain, or perhaps too slow, and information collecting may itself be costly. A fitness cost to developing and maintaining machinery to produce plastic responses and for making the responses is reasonable to expect as well. The most obvious way of introducing such costs into my models is to push down the two fitness curves (Fig. 3A, 3B). Hence, the phenotypes that can be adopted when making the plastic response only *tend* towards the respective optima because the more they move towards those extremes, the more it costs to the organism. The results of incorporating such constraints are displayed in Fig. 4.

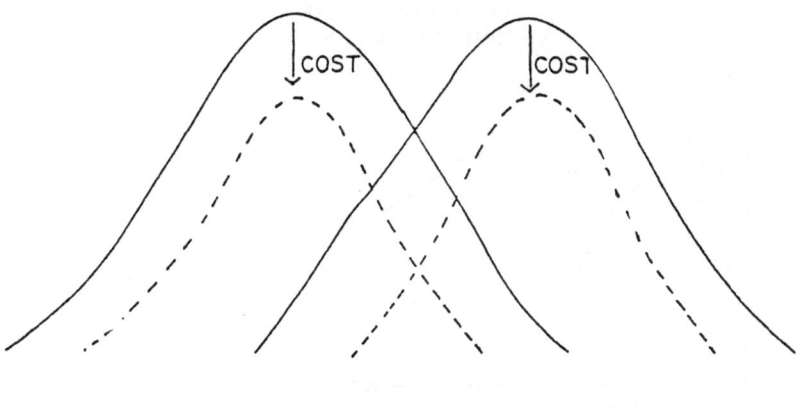

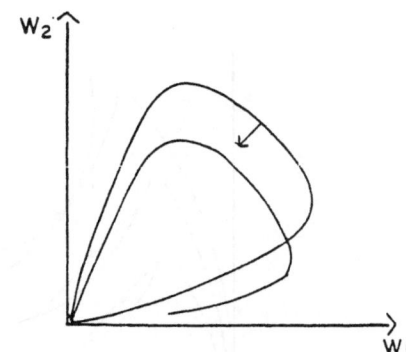

Figure 3. Costly plasticity.

Costs can be high or low (e.g., plasticity can be expensive or cheap). If it is *expensive*, the fitness curves, and therefore the fitness set, would have to be much pushed down before obtaining full plasticity (Fig. 4A). In this case, it is advantageous to be rigid when the opportunity border is concave. But when the latter is convex, environmental decipherability can make informed plasticity optimal, even when expensive (Fig. 4B). Convexity is not enough however, if the environment is inscrutable, because even with a convex border there is a selective advantage to rigidity over expensive ignorant plasticity (Fig. 4C). Of course, when plasticity is *cheap*, it will be selected for in most circumstances (Fig. 5), except when the environment is inscrutable and the opportunity border is concave.

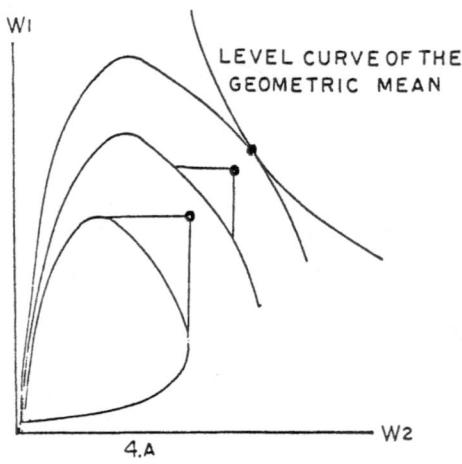

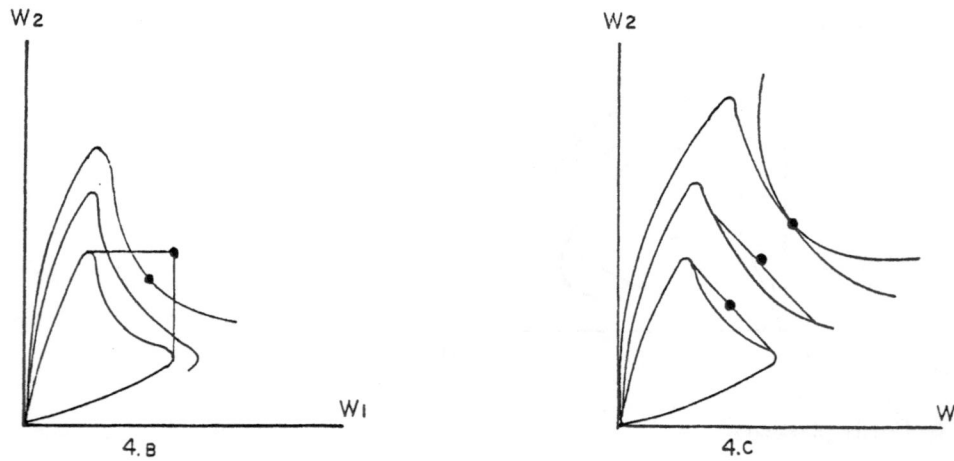

Figure 4. Expensive plasticity.

Imperfectly informed plasticity. Cues can deceive. A changing environment may only be decipherable for an organism if the cues available correlate with the upcoming circumstances. Such a correlation can be far from perfect, and to improve upon it may be difficult or costly. How can we introduce this 'informational imperfection' in our discussion? The cost aspect can be incorporated into the 'cost of plasticity' already treated, while the purely informational feature (e.g. reliability of cues) requires the use of conditional probability (Cohen 1967, León 1985). Suppose there are two signals, $\tilde{1}$

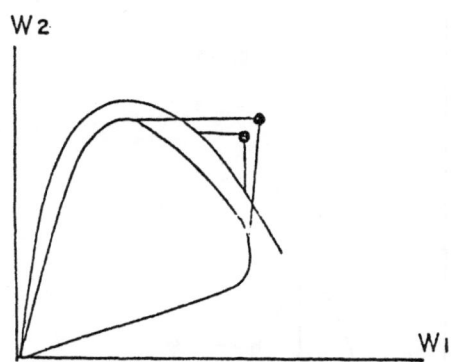 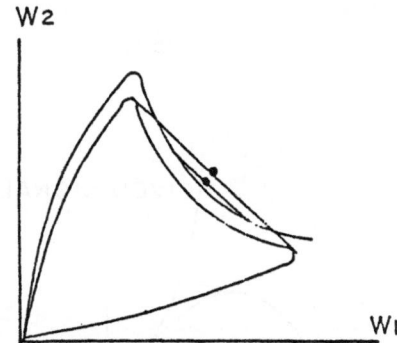

Figure 5. Cheap plasticity.

and $\widetilde{2}$, promising the advent of environmental states 1 and 2. Then we have a joint probability distribution specified by four values:

$$P(1,\widetilde{1}) \quad P(1,\widetilde{2}) \quad P(2,\widetilde{1}) \quad P(2,\widetilde{2})$$

In order to get some predictive value, we require:

$$P(1,\widetilde{1}) > P(1,\widetilde{2}) \qquad P(2,\widetilde{1}) < P(2,\widetilde{2})$$

Conditional probabilities are now defined as usual.

$$P(1|\widetilde{1}) = \frac{P(1,\widetilde{1})}{P(\widetilde{1})} = \frac{P(1,\widetilde{1})}{P(1,\widetilde{1}) + P(2,\widetilde{1})}$$

$$P(1|\widetilde{2}) = \frac{P(1,\widetilde{2})}{\Gamma(\widetilde{2})} = \frac{P(1,\widetilde{2})}{P(1,\widetilde{2}) + P(2,\widetilde{2})}$$

and so on. Of course, we have

$$P(1|\widetilde{1}) + P(2|\widetilde{1}) = 1$$
$$P(1|\widetilde{2}) + P(2|\widetilde{2}) = 1$$

The probabilities of occurrence of the environmental states can be expressed in terms of conditional probabilities:

$$q = P(1|\widetilde{1})P(\widetilde{1}) + P(1|\widetilde{2})P(\widetilde{2})$$
$$p = P(2|\widetilde{1})P(\widetilde{1}) + P(2|\widetilde{2})P(\widetilde{2})$$

Thus, the geometric mean

$$M = W_1^q W_2^p$$

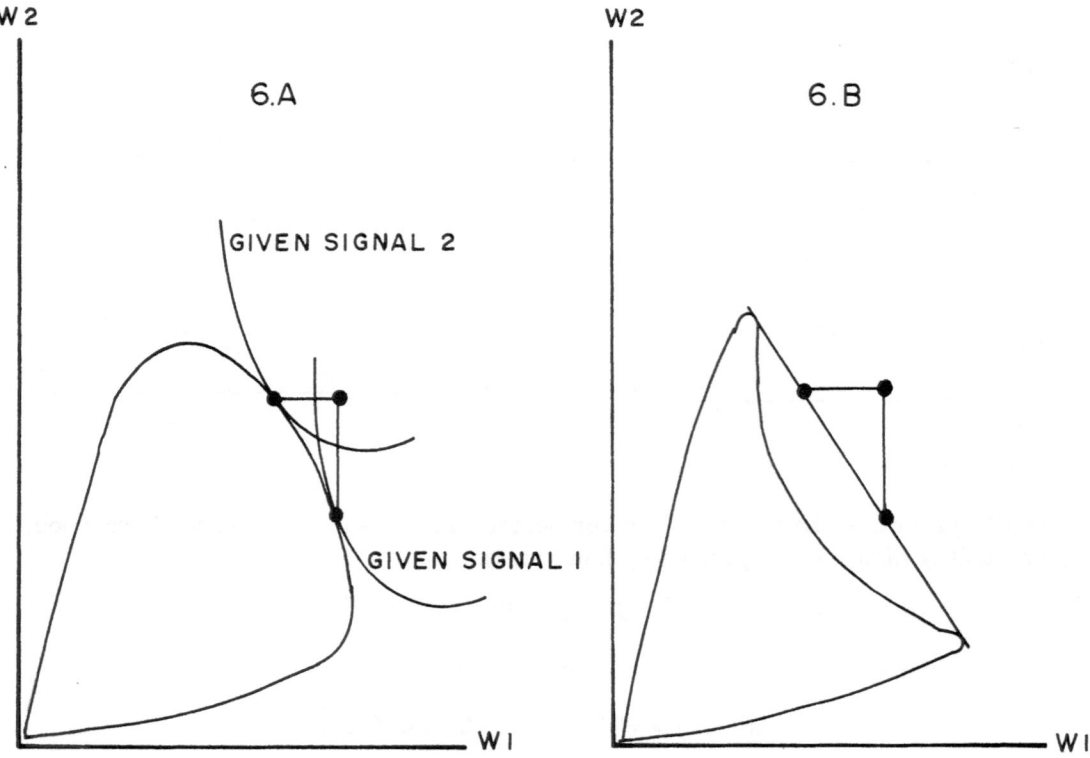

Figure 6. Imperfectly informed plasticity.

can be rewritten as a *geometric mean* of *conditional geometric means*:

$$M = M(\widetilde{1})^{P(\widetilde{1})} M(\widetilde{2})^{P(\widetilde{2})}$$

where:

$$M(\widetilde{1}) = W_1^{P(1|\widetilde{1})} W_2^{P(2|\widetilde{1})}$$
$$M(\widetilde{2}) = W_1^{P(1|\widetilde{2})} W_2^{P(2|\widetilde{2})}$$

Therefore, the optimization must proceed in two steps. First, we obtain the points in the opportunity border which maximize each conditional geometric mean. These stipulate the phenotypes that the organism must adopt (if the border is concave) or the 'coins' that must be used in a developmental decision (if the relevant border is a 'mixture line') according to the received cue. The second step is to combine these points in order to establish what level of the geometric mean is reached by the imperfectly informed strategist. Fig. 6 illustrates both steps.

These results indicate that the accuracy of obtainable information affects the degree of plasticity which can be advantageously attained. Entirely dependable cues permit phenotypic shifts to the respective optima as the circumstances change. When signals are occasionally false, plasticity must be reduced. The same must occur for the sort of 'meta-plasticity' which consist in changing "decision coins," that is, the range of usable 'probability switches' narrows. At the informational dead end in which decisions must be blind, plasticity or meta-plasticity are foregone because selection favors rigidity of phenotype or rigidity of 'decision device'.

Combined limitations.   These separate analyses so far have identified several limits to plasticity. The combined obstacles as well, would probably impede the evolution of flexibility if they acted in the same direction. These factors are high costs and low fidelity of cues. If plasticity is expensive or otherwise difficult, phenotypic change towards the different optima can only be attained at the expense of lowering fitness too much. The same effect is induced by the cost of information. Also, if cues are not reliable, it would be optimal not to commit to them entirely, instead achieving a phenotype short of the promised environmental condition. Maximizing conditional geometric means amounts to a reduction in optimal phenotypic range for both cost and unpredictability.

Discussion

The deliberate mathematical simplicity indulged throughout this essay has one advantage; it highlights the conceptual issues involved. Of course, several components of this preliminary synthesis were developed, or at least hinted at, by other authors. Levins briefly used fitness set analysis to discuss plasticity (Levins 1963, 1968a, 1968b). Schaffer (1974) turned to fitness sets in order to analyze his life-history model of the conventional "mixed versus monomorphic populations" dilemma. Conditional probabilities were introduced by Cohen (1967) and León (1985) to consider changes of germination fraction. Bradshaw (1965) has already underlined the relevance of cost to explain lack of plasticity, and Cohen (1967) suggested that information might be sometimes too costly as well but without incorporating such a cost into his models.

Besides collecting these notions in a framework, the main novelties here are: charging cost in a strategic analysis of plasticity in fluctuating environments, and using the conditional geometric mean for determining the phenotypes that should be adopted when information is not perfect.

Unfortunately allowing more than two environmental states precludes the benefits of a geometrical presentation. Instead optima may be derived perhaps by a Taylor expansion of $E(\ln W)$ as in our first formula (León 1983, Ritland 1983, Yoshimura and Clark 1991). Obviously, an explicit functional relationship between cost (loss of fitness) and degree of plasticity would be required for this multi-state problem. The analysis by Cohen (1966, 1967) of ignorant flexibility and informed meta-flexibility in seeds considers multiple environmental states, and even models assuming a continuous array of possible states have received some attention, as in cases of ignorant plasticity (León 1985, Bull 1987, Schultz 1991). Most of these studies however, do not include any costs to phenotypic plasticity.

Cost has been introduced explicitly into some models of plasticity in social environments (Fagen 1987), but without the effect of fluctuating environments; instead only frequency-dependent interactions in a physically constant environment are considered. Also Van Tienderen (1991) recently warned: "Both a flexible development capable of producing different phenotypes, and mechanisms for gathering and processing environmental information may be costly." Consequently, he introduced costs in his generalized version (1991) of the model that Via and Lande (1985) used to investigate evolution of a quantitative trait in a spatially heterogenous (but temporally unchanging) environment.

What has escaped our consideration up to this point is any difference of time scale between environmental shifts and plastic response, such as fine-grained fluctuations that are more rapid than the time step between reproductive episodes (which in the perennial model used herein has the same length as development time), or Levins's 'Epaminondas effect' (1968a), in which the plastic organism adopts a new phenotype only to find the environment changing again.

It is convenient to briefly discuss the relationship between categories or terms used in this article, and those found in the literature. Stearns (1989) recommends 'plasticity' as a general term for all kinds of environmentally induced phenotypic variation. Such umbrella usage has been followed here, extended even to random switches ('ignorant plasticity'). A 'reaction norm' is the continuous set of phenotypes that a genotype can produce according to the value of some environmental signal (Stearns 1989). This is what Schmalhausen (1949) called 'dependent development'. Since my models consider only two environmental states (two values of a signal), I have to leave out continuity. Many of the cases which I have presented, constitute examples of 'discrete reaction norms' because adding more and more environmental possibilities would fill in the set of phenotypes optimally produced. The norm will be 'inflexible' if the phenotype, once produced, cannot be reversed. In that case plasticity can only be observed as *different* individuals of the same genotype exhibiting different phenotypes. 'Flexibility', on the other hand, refers to the *same* individual changing phenotype (Stearns 1989). For instance, if the organisms portrayed in these models were 'annuals', only 'inflexible plasticity' would be possible, whereas 'flexibility' would be if they were perennials. Flexibility seems to correspond to the 'adaptive environmental modulation' of Smith-Gill (1983). When the phenotypes available are inherently discrete and disparate forms (e.g. dormancy vs. germination of seeds; castes of social insects; sexual vs. asexual reproduction in aphids, rotifers, cladocerans) we have a modality of plasticity called a 'developmental switch' by Levins (1963, 1968a), 'autonomous regulative development' by Schmalhausen (1949) and 'developmental conversion' by Smith-Gill (1983).

The essential conceptual stance adopted in this article precluded any attempt to invoke factual evidence, beyond incidental mention of some cases. The emphasis on cost (of plasticity and information) and on imperfect cues would require some empirical testing, of which I am largely unaware. Regardless of this, a large and growing body of literature on plasticity in many different traits and organisms is available, partial reviews of which are: Schlichting (1986) for plants; several papers in a special issue of Bioscience (1989, vol. 39, No. 7), by Stearns, Dodson, van Noordwijk and Schlichting; Kuiper and Kuiper (1988) for physiology; and Scheiner and Lyman (1989, 1991) for genetics.

References

Bradshaw, A.D. 1965. Evolutionary significance of phenotypic plasticity in plants. *Adv. Genet.* 13: 115–155.

Brown, J.S. & D.L. Venable. 1986. Evolutionary ecology of seed-bank annuals in temporally varying environments. *Amer. Nat.* 127: 31–47.

Bull, J.J. 1981. Evolution of environmental sex determination from phenotypic sex determination. *Heredity* 47: 173–184.

Bull, J.J. 1983. *The Evolution of Sex Determining Mechanisms.* Benjamin/Cummings, CA. USA.

Bull, J.J. 1987. Evolution of phenotypic variance. *Evolution* 41: 303–315.

Bulmer, M.G. 1984. Delayed germination of seeds: Cohen's model revisited. *Theor. Pop. Biol.* 26: 367–377.

Bulmer, M.G. 1985. Selection for iteroparity in a variable environment. *Amer. Nat.* 126: 63–71.

Caswell, H. 1983. Phenotypic plasticity in life-history traits: Demographic effects and evolutionary consequences. *Amer. Zool.* 23: 35–46.

Charnov, E.L. & J.J. Bull. 1977. When is sex environmentally determined? *Nature* 266: 35–46.

Charnov, E.L. & J.R. Krebs. 1973. On clutch size and fitness. *Ibis* 116: 217–219.

Cohen, D. 1966. Optimizing reproduction in a randomly varying environment. *J. Theoret. Biol.* 12: 119–129.

Cohen, D. 1967. Optimizing reproduction in a randomly varying environment when a correlation may exist between the conditions at the time a choice has to be made and the subsequent outcome. *J. Theoret. Biol.* 16: 1–14.

Cohen D. 1968. A general model of optimal reproduction in a randomly varying environment. *J. Ecol.* 56: 219–228.

Cohen, D. 1970. A theoretical model for the optimal timing of diapause. *Amer. Nat.* 104: 389–400.

Cooper, W.S. & R.H. Kaplan. 1982. Adaptive 'coin-flipping': a decision-theoretic examination of natural selection for random individual variation. *J. Theoret. Biol.* 94: 135–151.

Dodson, S. 1989. Predator-induced reaction norms. *Bioscience* 39: 447–452.

Ellner, S. (1985a). ESS germination strategies in randomly varying environments. I. Logistic-type models. *Theor. Pop. Biol.* 28: 50–79.

Ellner, S. (1985b). ESS germination strategies in randomly varying environments. II. Reciprocal yield laws. *Theor. Pop. Biol.* 28: 80–116.

Ellner, S. & C.A. Beuchat. 1984. A model of optimal thermoregulation during gestation by *Sceloporus jarrovi*, a live-bearing lizard. In: S.A. Levin & T.G. Hallam (eds) *Mathematical Ecology, Lecture Notes in Biomathematics* 54: 15–28. Springer-Verlag, Berlin.

Fagen, R. 1987. Phenotypic plasticity and social enironment. *Evol. Ecol.* 1: 263–271.

Frank, S.A. & M. Slatkin. 1990. Evolution in a variable environment. *Amer. Nat.* 136: 244–260.

Gillespie, J.H. 1973. Natural selection with varying selection coefficients: A haploid model. *Genet. Res.* 21: 115–120.

Gillespie, J.H. 1977. Natural selection for variance in offspring numbers: a new evolutionary principle. *Amer. Nat.* 111: 1010–1014.

Goodman, D. 1984. Risk spreading as an adaptive strategy in iteroparous life histories. *Theor. Pop. Biol.* 25: 1–20.

Gross, L.J. 1984. On the phenotypic plasticity of leaf photosynthetic capacity. In: S.A. Levin, & T.G. Hallam (eds.) *Mathematical Ecology, Lecture Notes in Biomathematics* 54: 1–14, Springer-Verlag. Berlin.

Haldane, J.B.S. & S.D. Jayakar. 1963. Polymorphism due to selection of varying direction. *J. Genet.* 58: 237–242.

Hanski, I. 1988. Four kinds of extra long diapause in insects: A review of the theory and observations. *Ann. Zool. Fennici* 25: 37– 53.

Kaplan, R.H. & W.S. Cooper. 1984. The evolution of developmental plasticity in reproductive characteristics: An application of the 'adaptive coin-flipping' principle. *Amer. Nat.* 123: 393–410.

Karlin, S. & U. Liberman. 1974. Random temporal variation in selection intensities: Case of large population size. *Theor. Pop. Biol.* 6: 355–382.

Kingsolver, J.G. & W.B. Watt. 1983. Thermoregulatory strategies in *Colias* butterflies: thermal stress and the limits to adaptation in temporally varying environments. *Amer. Nat.* 123: 393–410.

Klinkhamer, P.G.L., T.J. De Jong, T.J. Jong, J.A.J. Metz, & J. Val. 1987. Life history tactics of annual organisms: The joint effects of dispersal and delayed germination. *Theor. Pop. Biol.* 32: 127–156.

Kuiper, D. & P.J.C. Kuiper. 1988. Phenotypic plasticity in a physiological perspective. *Acta Oecologica/Oecol. Plant.* 9: 43–59.

León, J.A. 1983. Compensatory strategies of energy investment in uncertain environments. In: H.E. Freedman, C. Strobeck (eds.) *Population Biology, Lecture Notes in Biomathematics* 52: 85–90, Springer-Verlag, Berlin.

León, J.A. 1985. Germination strategies. In: P.J. Greenwood, P.H. Harvey, & M. Slatkin (eds.) *Evolution. Essays in honour of John Maynard Smith.* Cambridge University Press. Cambridge, pp. 129–142.

Levin, S.A., Cohen, D. & A. Hastings. 1984. Dispersal strategies in patchy environments. *Theor. Pop. Biol.* 26: 165–191.

Levins, R. 1963. Theory of fitness in a heterogeneous environment II. Developmental flexibility and niche selection. *Amer. Nat.* 98: 75–90.

Levins, R. 1968a. *Evolution in Changing Environments.* Princeton University Press. Princeton.

Levins, R. 1968b. Evolutionary consequences of flexibility. In: R.C. Lewontin (ed.) *Population Biology and Evolution.* Syracuse University Press. Syracuse.

Lively, C.M. 1986. Canalization versus developmental conversion in a spatially variable environment. *Amer. Nat.* 128: 561–572.

Lively, C.M. 1987. Facultative parthenogenesis and sex-ratio evolution. *Evol. Ecol.* 1: 197–200.

McGinley, M.A., D.H. Temme,& M.A. Geber. 1987. Parental investment in offspring in variable environments. *Amer. Nat.* 130: 370–398.

Metz, J.A.J., T.J. De Jong, T.J. Jong, & P.G.L. Klinkhamer. 1983. What are the advantages of dispersing; a paper by Kuno explained and extended. *Oecologia* 57: 166–169.

Oster, G. & E.O. Wilson. 1978. *Caste and Ecology in the Social Insects*. Princeton University Press. Princeton.

Real, L.A. 1980. Fitness, uncertainty, and the role of diversification in evolution and behavior. *Amer. Nat.* 115: 623–638.

Ritland, K. 1983. The joint evolution of seed dormancy and flowering time in annual plants living in variable environments. *Theor. Pop. Biol.* 24: 213–243.

Russell, B. 1945. *A History of Western Philosophy*. Simon & Schuster. New York.

Schaffer, W.M. 1974. Optimal reproductive effort in fluctuating environments. *Amer. Nat.* 108: 783–790.

Scheiner, S.M. & R.F. Lyman. 1989. The genetics of phenotypic plasticity I. Heritability. *J. Evol. Biol.* 2: 95–107.

Scheiner, S.M. & R.F. Lyman. 1991. The genetics of phenotypic plasticity II. Response to selection. *J. Evol. Biol.* 4: 3–50.

Schlichting, C.D. 1986. The evolution of phenotypic plasticity in plants. *Ann. Rev. Ecol. Syst.* 17: 667–693.

Schlichting, C.D. 1989. Genotypic integration and environmental change. *Bioscience* 39: 460–464.

Schmalhausen, I.I. 1949. *Factors of Evolution*. Blakiston. Philadelphia.

Schultz, D.L. 1991. Parental investment in temporally varying environments. *Evol. Ecol.* 5: 415–427.

Smith, C.C. & S.D. Fretwell. 1974. The optimal balance between size and number of offspring. *Amer. Nat.* 108: 499–506.

Smith-Gill, S.J. 1983. Developmental plasticity: Developmental conversion versus phenotypic modulation. *Amer. Zool.* 23: 47–55.

Stearns, S.C. 1989. The evolutionary significance of phenotypic plasticity. *Bioscience* 39: 436–445.

Stearns, S.C. & J.C. Koella. 1986. The evolution of phenotypic plasticity in life-history traits: predictions of reaction norms for age and size at maturity. *Evolution* 40: 893–913.

Tuljapurkar, S.D. 1982. Population dynamics in variable environments III. Evolutionary dynamics of r-selection. *Theor. Pop. Biol.* 21: 141–165.

Tuljapurkar, S.D. 1989. An Uncertain life: Demography in random environments. *Theor. Pop. Biol.* 35: 227–294.

van Noordwijk, A.J. 1989. Reaction norms in genetical ecology. *Bioscience* 39: 453–458.

Van Tienderen, P.H. 1991. Evolution of generalists and specialists in spatially heterogeneous environments. *Evolution* 45: 1317–1331.

Venable, D.L. & J.S. Brown. 1988. The selective interactions of dispersal, dormancy and seed size as adaptations for reducing risk in variable environments. *Amer. Nat.* 131: 360–384.

Venable, D.L. & L. Lawlor. 1980. Delayed germination and dispersal in desert annuals: escape in space and time. *Oecologia* 46: 272–82.

Via, S. & R. Lande. 1985. Genotype-environment interaction and the evolution of phenotypic plasticity. *Evolution* 39: 505–522.

Yodzis, P. 1989. *Introduction to Theoretical Ecology*. Harper & Row.New York.

Yoshimura, J. & C.W. Clark. 1991. Individual adaptations in stochastic environments. *Evol. Ecol.* 5: 173–192.

OPTIMIZATION AND ESS ANALYSIS

FOR POPULATIONS IN STOCHASTIC ENVIRONMENTS

Colin W. Clark
Institute of Applied Mathematics
The University of British Columbia
Vancouver, B.C. V6T 1Z2

and

Jin Yoshimura
Department of Zoology
Duke University
Durham, NC 27708-0325

Abstract. The interplay of demographic and environmental variance is considered in a behavioral context involving foraging in the presence of uncertain, fluctuating risk of predation. The evolutionary contest between two strategies – high risk and high fecundity versus low risk and low fecundity – is considered. Often the risky strategy will be dominant when it is sufficiently abundant in the population. Demographic (sampling) variance, however, implies that either strategy may be at a disadvantage when rare. In a phenotypic model the stable strategy may be determined by initial conditions and by chance. In a genetic model, polymorphism may be maintained if the risky strategy is heterozygotic.

The use of mathematical models is generally accepted as an indispensable component of the study of evolutionary processes. The scope of mathematical modeling, and the choice of mathematical technique, continue to widen as more details of the evolutionary system come under scrutiny. In this article we discuss optimization models, and their extension to ESS models, in relation to stochastic environments and their effects on the evolution of life-history and behavioral strategies.

Optimization models

The use of optimization models in biology has sometimes been misconstrued as representing an uncritical belief in the perfection of nature (e.g. Gould and Lewontin, 1979). Only the naive believe that any particular mathematical model can provide more than a partial view of the complexity inherent in any biological system (Kingsland, 1985; Fagerström, 1987). Any model, by definition, is a simplified representation of some natural system or process. Models are useful to the extent to which they help us to understand the logical consequences of structures and processes (whether known or hypothesized), especially in situations where these consequences are not obvious beforehand. Optimization and game-theoretic (ESS) models have been useful, indeed instrumental, in understanding the evolution of many behavioral and life-history traits. The status of optimization models in biology has been discussed in Dupré (1987) and by Mitchell and Valone (1990) and Parker and Maynard Smith (1990).

Early optimization models of life-history traits and behavioral decisions typically concerned only single traits or decisions, occurring under given environmental conditions. In reality, multiple traits often interact, involving various tradeoffs in terms of fitness consequences. Behavioral decisions occur sequentially, with each decision affecting later opportunities. Environments are patchy, and fluctuate randomly over time. Finally, fitness consequences often depend on the behavior, and the abundance of other organisms.

Several developments in modeling techniques have resulted from the need to make optimization analysis more general and useful. Game-theoretic (ESS) models have been used to study density-dependent effects and interactions between organisms (Maynard Smith, 1982). Evolution in stochastic environments has been modeled from an optimization perspective by Levins (1968), Lewontin and Cohen (1969), and others. Dynamic state variable optimization models have been used to study sequential decisions and multiple tradeoffs in stochastic environments (Mangel and Clark, 1988).

Stochastic environments

In discussing the role of stochasticity in optimization models, we immediately encounter an important problem: such models require an optimization objective, usually identified with "fitness" in some way. This is straightforward in a deterministic model – fitness is simply defined as the number of viable offspring of an individual (or the dominant eigenvalue in an age-structured life-history model). But in a stochastic environment the number of offspring per individual is a random variable. What is it that is then maximized, if anything?

The simplest assumption is that natural selection maximizes average or expected individual reproduction. For example, the optimal clutch size for a given species of bird, in a given stochastic environment, might be assumed to be that which maximizes the average number of fledgelings per individual.

What is the justification for such an assumption? Compare, for example, two sharply contrasting clutch-size strategies. Strategy 1, the "risky" strategy, produces five fledgelings if the parent encounters a good environment, but none in a bad environment. Strategy 2, the "conservative" strategy, produces two fledgelings in all environments. (The parent cannot predict which environment will be encountered). If p denotes the probability of encountering a bad environment, then the expected fitness of the risky strategy is $5(1-p)$. Thus the risky strategy is superior (according to the expected fitness criterion) provided that $5(1-p) > 2$, i.e. $p < 0.6$. Is this a reasonable criterion?

To answer this question let us first make the additional simplifying assumption that the species in question only reproduces once. Could a gene coding for the risky strategy (either 5 or 0 offspring) invade and persist in a population of conservative strategists? Certainly not, at least if the environmental conditions affect all individuals in the same way – for then a single bad season would result in the complete elimination of the risky morph. This remains true no matter how rarely the disastrous environmental conditions occur.

The foregoing example, deliberately extreme, already shows that the maximization of average or expected reproductive output is not always an appropriate criterion for understanding natural selection in a stochastic environment. For models of this kind, it has long been recognized that the proper criterion is related to the geometric mean of reproductive success, rather than the arithmetic mean – see Levins (1968), Lewontin and Cohen (1969).

For a model with n environmental states i, occurring with probability p_i and yielding rewards f_i the arithmetic mean is $\sum_i p_i f_i$ whereas the geometric mean is $\prod_i f_i^{p_i}$. In the above model the conservative strategy has arithmetic and geometric mean both equal to 2, whereas the risky strategy has arithmetic mean $5(1-p)$ and geometric mean 0. Thus the risky strategy is always inferior to the conservative strategy, according to the geometric mean criterion.

But now suppose that different members of the population encounter different environmental conditions *independently*. The probability that the risky morph will become extinct in one generation is then p^n where p is the probability that an individual clutch encounters an unfavorable environment, and n is the size of the risky morph population. If n is reasonably large this probability becomes vanishingly small. For example, if $p = 0.5$ (half the clutches fail) and $n = 100$, then the probability of extinction in one generation is about one in 10^{30}, and the expected persistence time of the risky morph is at least 10^{30} generations. If the risky strategy has greater average fecundity than the conservative strategy it will tend to take over, once it has initially established itself, long before it reaches extinction.

But would the risky strategy succeed in becoming established, starting with, say, a single mutant individual? Obviously this depends on chance: if several good seasons occur consecutively the risky morph might survive and become dominant; but an early sequence of bad seasons might destroy the risky morph.

The point of this simple model is to show that the standard paradigm, whereby strategies that have higher reproductive success "on average" are invariably selected over other strategies, is in need of revision for populations living in stochastic environments. The outcome of an evolutionary contest may depend on chance, in a way that is strongly dependent on the way that stochastic events affect different individuals in the population. In particular, a risky but highly productive strategy may be inferior when rare, but dominant once it fortuitously becomes sufficiently abundant (Clark and Yoshimura, 1993). In practice, "environmental" stochasticity (which affects all individuals in the population in the same way) and "demographic" stochasticity (which affects individuals independently) may often occur together.

Interplay of demographic and environmental variance

The expression

$$\mu - \frac{1}{2} \frac{\sigma^2}{\mu} \tag{1}$$

is a well known approximation for the geometric mean, where μ denotes the arithmetic mean, and σ^2 the variance, in per capita reproduction of a given population. This simple expression indicates the significance of variance discounting in stochastic environments: given two reproductive strategies with the same arithmetic mean (μ), the strategy with the lower variance σ^2 will have the higher geometric mean, and will tend to persist (Gillespie, 1974). More generally, Eq. (1) shows that a tradeoff exists between mean and variance; reproductive strategies that sacrifice mean reproductive output in order to reduce variance have been called "bet-hedging" strategies (Slatkin, 1974; Lacey et al., 1983; Seger and Brockmann, 1987). For example, Boyce and Perrins (1987) suggest that bet-hedging may account for the observed discrepancy between observed and predicted clutch sizes in great tits.

The variance σ^2 in Eq. (1) may result from environmental or demographic stochasticity, or both.

Specifically, let $w = w(t)$ be a random variable representing the environmental state in year t. Suppose that the ith member of a given population has reproductive output $X_i(w)$, where the $X_i(w)$ are i.i.d. random variables conditional on w, with (conditional) variance $\sigma_X^2(w)$. Let $\mu(w) = \Sigma_i X_i(w)/N$ be the population mean per-capita reproduction. It is then easy to show that overall variance is given by

$$\sigma^2 = \frac{1}{N} E_w(\sigma_X^2(w)) + \sigma_Y^2 \tag{2}$$

where N denotes total population size, E_w is the mathematical expectation with respect to w, and

$$\sigma_Y^2 = E_w(\mu(w) - \bar{\mu})^2 \tag{3}$$

is the environmental variance (Yoshimura and Clark, 1991). Equation (2) partitions total variance among demographic variance and environmental variance. For large populations environmental variance dominates but for small populations individual variance may be important. A strongly fluctuating environment may induce fluctuations in population size N, so that the relative importance of environmental and demographic variance in reproductive strategies will alternate over time. (Since N is also a random variable, Eq. (2) is itself only an approximation. Nevertheless it does indicate the relative importance of environmental and demographic variance, and the role of population size.)

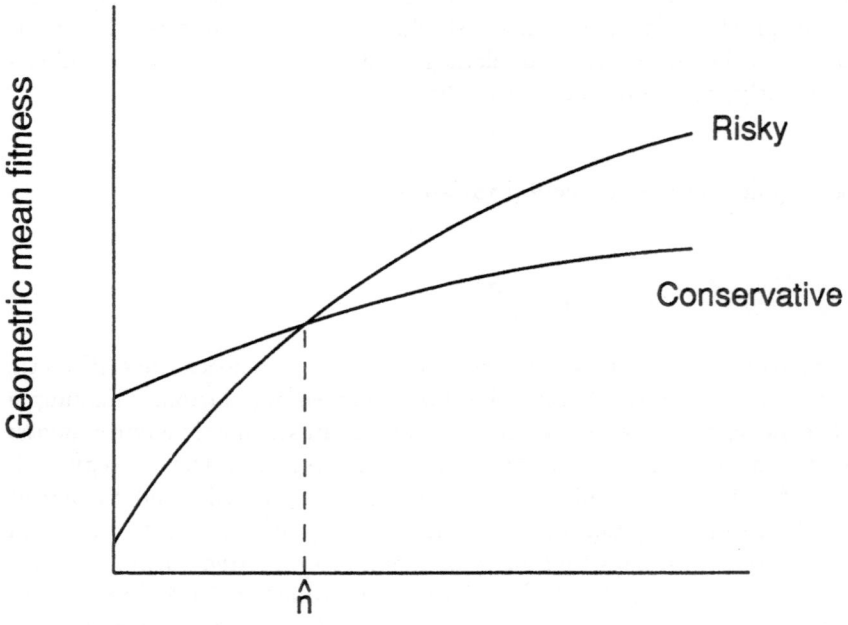

Figure 1. Geometric mean fitness as a function of morph population sizes n_i for conservative and risky morphs. For the case shown here, the two curves cross over at $n_i = \hat{n}$; when both $n_i < \hat{n}$ the conservative morph is superior, whereas the risky morph is superior when both $n_i > \hat{n}$. When one morph is above and the other below $\hat{n}$, the more abundant morph is superior and relatively immune from invasion by the other morph.

Consider again an evolutionary competition between two reproductive strategies, one risky and the other conservative. For example, in Clark and Yoshimura (1993) we model foraging in the presence of predation risk, assuming that the annual abundance of predators is a random variable. The risky foraging strategy accepts a high risk of predation in order to obtain a high rate of food retrieval; the conservative strategy does the reverse. Denote by $q_i(w)$ the risk of predation, per season, under strategy i, where $i = 1$ is the risky strategy and $i = 2$ is the conservative strategy. Thus

$$q_1(w) > q_2(w) \qquad \text{for all } w$$

The risky strategist, if she survives, attains the greater fecundity:

$$f_1 > f_2$$

Figure 1 shows the geometric mean fitness curves, as functions of the two strategy-morph population sizes n_i, for this model. In the situation depicted (other relative positions of the two curves are possible), the risky morph is inferior at low population numbers, but becomes superior when sufficiently abundant. The risky strategy is *individually* risky, but if n_1 is large enough the individual (demographic) risks become less important. As a rare mutant, the risky strategy would be unlikely to invade an established population of the conservative morph. On the other hand, if the risky morph did become established, the conservative morph would be unlikely to reinvade. Provided that environmental variance is not too large, the initially established morph is the evolutionarily stable type.

In a highly variable environment both populations n_i could fluctuate drastically. Following a bottleneck event either morph could win out and become established. In Clark and Yoshimura (1993) we describe the results of various simulation experiments with this model. Stochastic alternation in abundance of the two morphs was one possible outcome.

Note that it does not make sense to use a linear (density independent) population model here, because of the influence of population numbers n_i. Our simulation models therefore incorporated a saturation component, acting on the combined population. As a result, one or the other of the two morphs quickly became extinct in most of our simulations. If we prevented extinction from occurring, by always maintaining a remnant population of each type, the result (in a strongly fluctuating environment) was a stochastic alternation in abundance of the two types (Clark and Yoshimura, 1993, Fig. 4). But how might such remnant populations be maintained in nature? One possibility is geographical heterogeneity: different regions would experience population crashes at different times, and migration would re-establish types that had become locally extinct. A second possibility is that the genetic system might maintain sufficient variability; a version of our model incorporating Mendelian genetics is discussed in the Appendix.

Conclusions

Optimization and game-theoretic (ESS) models have been widely used in deterministic form. Most biological organisms, however, live in fluctuating, unpredictable environments, and many have evolved adaptations to deal with this stochasticity. Stochastic environments often induce adaptive variation in life history and behavior; these biologically important variations can obviously not be studied by means of deterministic (or long-term averaged) models. On the basis of such models, observed variations can only be interpreted as meaningless statistical noise about a predicted value.

In this paper we have discussed the interplay of environmental and demographic stochasticity, the distinction being that environmental stochasticity affects all members of a population in a tightly correlated manner, whereas demographic stochasticity acts independently upon different individuals. Here the time scale of environmental fluctuations is assumed to correspond to a single generation; this case is referred to as a "coarse-grained" environment (e.g Levins, 1968). In a "fine-grained" environment, on the other hand, fluctuations occur on a time scale much shorter than the intergenerational period. Because individuals sample a fine-grained environment many times within a given reproductive cycle, stochastic effects are largely averaged out at the intergenerational scale. In this case the arithmetic mean remains a useful construct for defining fitness.

Short-term environmental fluctuations may result in short-term variations in behavior, not only as immediate responses to current environmental conditions but also as reactions to variations in the organism's internal (physiological) state, which is itself influenced by past environmental events and responses. Behavioral adaptations may also prepare for expected future environmental conditions, as cued by present and recent past events. Behavioral adaptations and phenotypic variation in behavior have been successfully studied by means of stochastic dynamic state-variable models (Mangel and Clark, 1988). This approach has been used to develop novel adaptationist interpretations of such field observations as diel vertical migration of planktivores and zooplankton (Clark and Levy, 1988), variance in clutch size of parasitic insects (Mangel, 1987), the timing of fledging in birds (Ydenberg, 1989), and many others.

In these stochastic dynamic models fitness is defined in terms of expected lifetime reproduction, where "expected" refers of course to the arithmetic mean. This is reasonable provided that intergenerational variance is not important – in other words provided that the stochasticity is fine-grained.

It is unlikely that any simple measure of fitness is optimized in a coarse-grained stochastic environment. In simulation models (nonoverlapping generations), we have found that the geometric mean concept often provides a useful guide for predicting qualitative results, but in many cases no single strategy can be identified as being uniquely optimal, or evolutionarily stable. The winning strategy in a simple simulation model is often determined by chance, depending on initial conditions. With additional, realistic components of complexity (age structure, spatial heterogeneity, genetics, etc.), coexistence of multiple life history and behavioral strategies would seem to be the most likely outcome. Stochastic evolutionary game theory provides a useful approach to this question, but has only recently begun to be developed (e.g. Levin et al., 1984; Hairston and Munns, 1984; Ellner, 1985; Cohen and Levin, 1991; Ludwig and Levin, 1991; Clark and Yoshimura, 1993).

References

Boyce, M. and C.M. Perrins. 1987. Optimizing great tit clutch size in a fluctuating environment, *Ecology* 68:142–153.

Clark, C.W. and D.A. Levy. 1988. Diel vertical migrations by juvenile sockeye salmon and the antipredation window, *Amer. Nat.* 131: 271–290.

Clark, C.W., and J. Yoshimura. 1993. Behavioral responses to variations in population size: a stochastic evolutionary game, *Behav. Ecol.* (in press).

Cohen, D., and S. Levin. 1991. Dispersal in patchy environments: the effects of temporal and spatial structure, *Theoret. Pop. Biol.* 39: 63-99.

Dupré, J. (Ed.) 1987. *The Latest on the Best: Essays on Evolution and Optimality.* MIT press, Cambridge, MA.

Ellner, S. 1985. ESS germination strategies in randomly varying environments. I. Logistic-type models, *Theoret. Pop. Biol.* 28: 50–79.

Fagerström, T. 1987. On theory, data and mathematics in ecology, *Oikos* 50: 258–261.

Gillespie, J.H. 1974. Natural selection for within-generation variance in offspring number, *Genetics* 76: 601–606.

Gould, S.J. and R.C. Lewontin. 1979. The spandrels of San Marco and the Panglossian paradigm: a critique of the adaptational programme, *Proc. Roy. Soc. London. B.* 205: 581–598.

Hairston, N.G., and W.R. Munns, Jr. 1984. The timing of copepod diapause as an evolutionarily stable strategy, *Amer. Nat.* 123: 733–751.

Kingsland, S. 1985. *Modeling Nature.* University of Chicago Press, Chicago, IL.

Lacey, E.P., L.A. Real, J. Antonovics, and D.G. Heckel. 1983. Variance models in the study of life histories, *Amer. Nat.* 122: 114–131.

Levin, S.A., A. Hastings and D. Cohen. 1984. Dispersal strategies in patchy environments, *Theoret. Pop. Biol.* 26 165–191.

Levins, R. 1968. *Evolution in Changing Environments.* Princeton University Press, Princeton, N.J.

Lewontin, R.C. and D. Cohen. 1969. On population growth in a randomly varying environment, *Proc. Nat. Acad. Sci. USA* 62: 1056–1060.

Ludwig, D.A. and S.A. Levin. 1991. Evolutionary stability of plant communities and the maintenance of multiple dispersal types, *Theoret. Pop. Biol.* 40: 285–307.

Mangel, M. 1987. Oviposition site selection and clutch size in insects, *J. Math. Biol.* 25: 1-22.

Mangel, M. and C.W. Clark. 1988. *Dynamic Modeling in Behavioral Ecology.* Princeton University Press, Princeton, NJ.

Maynard Smith, J. 1982. *Evolution and the Theory of Games.* Cambridge University Press, Cambridge.

Mitchell, W.A. and T.J. Valone. 1990. The optimization research program: Studying adaptations by their function, *Quart. Rev. Biol.* 65: 43–52.

Parker, G.A. and J. Maynard Smith. 1990. Optimality theory in evolutionary biology, *Nature* 348: 27–33.

Seger, J. and J. Brockmann. 1987. What is bet-hedging? *Oxford Surveys in Evolutionary Biology* 4:182-211.

Slatkin, M. 1974. Hedging one's evolutionary bets, *Nature* 250: 704–705.

Ydenberg, R.C. 1989. Growth-mortality trade offs, parent-offspring conflict, and the evolution of juvenile life histories in the avian family, Alcidae, *Ecology* 70: 1496–1508.

Yoshimura, J. and C.W. Clark. 1991. Individual adaptations in stochastic environments, *Evol. Ecol.* 5: 173–192.

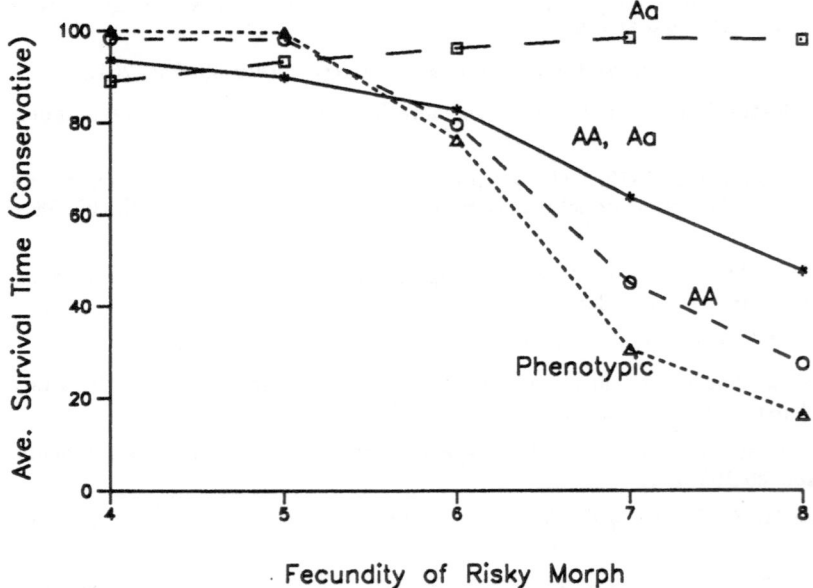

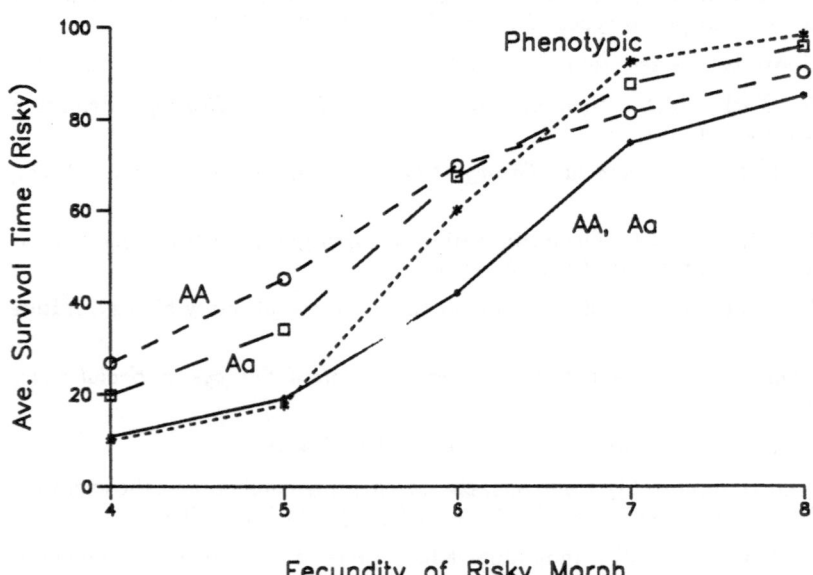

Figure 2. Average survival times (over 100 generations) for conservative and risky phenotypes, from Mendelian model. Horizontal axis represents fecundity of risky morph; conservative morph fecundity equals 3. Case (1): risky phenotype = AA, Aa (—); case (2): risky phenotype = AA (short dashes); case (3): risky phenotype = Aa (long dashes). The results from a nongenetic version of the model are also plotted (···). Note that, when fecundity of the risky morph is high, polymorphism is strongly maintained provided that the risky type is heterozygotic (curve Aa). Parameter values: survival probabilities: 0.6, 0.9 for conservative type, 0.2, 0.7 for risky type, in bad or good environment respectively; carrying capacity = 100 individuals of both types; initial population = 40; initial gene frequency = 0.5; probability of bad year = 0.5.

Appendix: A Mendelian model

We considered a single-locus, two-allele model of standard type, without mutation or genetic drift, but with stochastic fitness coefficients as in the phenotypic model discussed earlier. We considered three cases:

(1) Risky type dominant: genotypes AA, Aa are risky, aa conservative.
(2) Risky type recessive: genotype AA is risky, Aa, aa conservative.
(3) Risky type heterozygotic: genotype Aa is risky, AA, aa conservative.

The assumed sequence of life-history events is that young zygotes forage under risk of predation (exactly as in the purely phenotypic model), surviving adults undergo panmictic breeding, producing offspring whose total number is limited by saturation, leading to the next generation.

The results of our simulations, exemplified in Fig. 2, are readily understood in terms of the geometric-mean fitness diagram of Fig. 1. In case 3 (risky type heterozygotic), for example, genetic diversity is maintained by heterozygote superiority, when both alleles are abundant. When the total population is small, one or the other allele tends towards fixation, but this situation is usually ephemeral, so that neither allele is eliminated. Of course, if the risky type (heterozygote) is inferior at all population levels (i.e. f_1 small – see Fig. 2b), then one allele tends towards fixation, so that the heterozygote (risky phenotype) disappears over time.

It is generally understood that heterozygote superiority is likely to evolve in stochastic environments. Our analysis extends this prediction to the case of density-dependent competition between risky and conservative strategies. We expect that the risky phenotype will often be the heterozygote in this situation.

In our models we have assumed that the carrying capacity K is larger than the population level $\hat{n}$ at which the two fitness curves cross over (Fig. 1); the opposite case is fairly uninteresting. However, in a stochastic environment it might be the case that K fluctuates. Certain species, for example, face harsh environments most of the time, with only occasionally favorable conditions. If so, the conservative strategy is usually superior to the risky one, but the risky strategy allows for more rapid population growth under favorable conditions. Under these circumstances the conservative phenotype may be heterozygotic, with both homozygotes being risky (this would be case (4) in our classification).

ARE VARIABLE ENVIRONMENTS STOCHASTIC?

A REVIEW OF METHODS

TO QUANTIFY ENVIRONMENTAL PREDICTABILITY

Steven R. Beissinger and James P. Gibbs
School of Forestry & Environmental Studies
Yale University
205 Prospect St.
New Haven, CT 06511

Abstract. We review some methods for quantifying environmental predictability and discuss their advantages and disadvantages. Colwell's (1974) index can be used to give an absolute measure of predictability, which can be partitioned into constancy and contingency (seasonality), and is applied to time series of categorical data. Spectral analysis is useful for measuring periodicity or seasonality of environments, and requires time series of continuous data. We compare the behavior of these measures for 15 wetland sites in tropical and subtropical portions of North and South America. Estimates of seasonality and predictability based on spectral densities were related positively to estimates of predictability and contingency, and to some extent negatively with constancy, derived from Colwell's index. These methods performed better than simple statistical measures of variation such as the standard deviation or coefficient of variation. Studies seeking to characterize environmental variation and explore its implications for life-history evolution should employ these time series approaches.

Introduction

Perhaps no concept in ecology is more misused than environmental predictability. Environments have been called unpredictable when they exhibit fluctuations or variations in conditions, or when disturbances or catastrophes occur (e.g., Palumbi 1984, Eckert

1987, Wrege and Emlen 1991). However, not all environmental fluctuations are unpredictable or stochastic. In fact, many environmental fluctuations of great magnitude are seasonal and occur in a very predictable manner.

The theme of this volume is understanding adaptation and selection under stochastically fluctuating environments. Implicit in this theme is the assumption that environments vary stochastically and that we have some understanding of the nature of environmental variability. In fact, our understanding of environments and how they vary is rudimentary. While many studies have quantified environmental variation, few have examined whether such variation is predictable. In this article, we review some methods for quantifying environmental predictability in hopes of stimulating more researchers to characterize the environments in which they are studying adaptation.

Natural selection acts on organisms in particular environments. Broadly defined, the environment includes not only the physical or climatic characteristics of a habitat, but also the social setting that an organism inhabits. For example, for species that live in groups, the most important aspect of the environment may be the age, sex, and size of other members of its group, since these characteristics often affect whether and when an individual may ascend to a breeding position (Zack and Rabenold 1989). The predictability of social environments is usually difficult to determine because time series of sufficient length for social characteristics are rare. As more long-term studies are completed (Newton 1990), perhaps appropriate data sets will become available to determine the predictability of some social environments.

On the other hand, physical environments are usually well documented because records of climatic conditions have been carefully compiled in many localities around the world for decades and even for centuries in some locales. Long-term records of monthly or daily precipitation, air temperature, sea temperature, barometric pressure, solar radiation, and relative humidity are available on computer tape for many locations in North America and for selected locations on every continent from the U.S. National Climatic Data Center in Asheville, North Carolina. Although climatic conditions may often vary greatly at the local level in some landscapes, it is likely that data from nearby weather stations would covary positively with the particular locale of study. This should be especially true for long-term trends or periodicities of environments.

Predictability, constancy, and contingency

Colwell (1974) developed simple measures of environmental variation that depend upon classifying environments into categories or "states." For example, one might be interested in determining the predictability of flowering or fruiting in a tropical tree (Colwell 1974) or the occurrence of rainfall in deserts (Low 1979). To a bird or ant foraging upon the fruits of a tree, environmental states are clearly defined as "flowering," "fruiting," or "neither," whereas the states "not rain enough to initiate grass growth," "enough rain to initiate grass growth," and "enough rain to initiate and continue grass growth" describe the way a herbivore might view its chances of finding food in a desert based on Slater's (1962) concept of 'effective rainfall'.

The predictability (P) of an environment can be considered the sum of its constancy (C) and contingency (M) or seasonality of occurrence (Colwell 1974). Predictability can result from constancy, i.e. the state of the environment never changes. But predictability may also result entirely from contingency, i.e. the occurrence of a state is dependent on the previous occurrence of a different state. Under such conditions, environments may be highly variable but can be very predictable.

Colwell's (1974) index of environmental predictability varies from 0 to 1 and uses information theory to compute the uncertainty of an environment with respect to time, state and their interaction. A matrix is constructed using time categories (e.g., seasons or months of the year) as columns and environmental states (s) as rows. Let N_{ij} equal the number of times that the environment was in state i at time j. Column totals are defined as X_j , row totals as Y_i , and the grand total is Z. Then predictability is:

$$P = 1 - \frac{H(XY) - H(X)}{\log s} \qquad (1)$$

where

$$H(X) = -\Sigma_j \frac{X_j}{Z} \log \frac{X_j}{Z}$$

and

$$H(XY) = -\Sigma\Sigma \frac{N_{ij}}{Z} \log \frac{N_{ij}}{Z}$$

P can be decomposed into:

$$C = 1 - \frac{H(Y)}{\log s} \qquad (2)$$

where

$$H(Y) = -\Sigma_i \frac{Y_i}{Z} \log \frac{Y_i}{Z}$$

and

$$M = \frac{H(X) + H(Y) - H(XY)}{\log s} \qquad (3)$$

See Colwell (1974), Stearns (1981) and Gan et al. (1991) for details on the derivation and calculation of the index. A G-test (Sokal and Rolf 1981) may be used to determine if P, C, and M are significantly different from zero (i.e, stochastic) using critical values given in Stearns (1981).

This index has the virtue of yielding an absolute scale for environmental predictability that facilitates comparisons among habitats, locations, or time periods. It also allows predictability to be decomposed into two components with different significance for adaptive strategies – constancy and contingency. Constant environments are thought to facilitate adaptation because selection can act continually on organisms (Futuyma 1986). Environments that vary seasonally or contingently may be more difficult for selection to track, but may facilitate adaptive strategies if organisms can use reliable cues to predict environmental changes (Low 1978). For example, seasonal patterns of rainfall may be predictable from changes in daylength or temperature.

Stearns (1981) investigated the statistical properties of Colwell's index using simulated time-series data. He found that both C and M were powerful, and that power increased with the strength of the signal. But whereas C was powerful with both small (< 30) and large (> 300) sample sizes, the number of Type I errors rose for M with small sample sizes. As expected, M and C took on values approaching 0 and 1, respectively, for constant time series, whereas values of M and C approached 1 and 0, respectively, for strongly periodic time series. Colwell's indices could detect long-term periodicities in data sets by varying the length (and number) of time periods to determine where M and P were maximized. But Fourier analysis on metric data sets detected signals at lower signal-to-noise ratios and at smaller sample sizes than Colwell's index calculated by categorizing the same data sets.

Several analyses have shown that Colwell's index is preferable to simple measures of environmental variation. No strong correlations were found between Colwell's index and means, variances, or coefficients of variation (Low 1978, Gan et al. 1991). This reinforces the notion that the variance of an environmental attribute may not be related to its predictability. In other words, highly variable environments are not necessarily stochastic environments.

Colwell's index has been employed to examine the predictability of water levels for a snail-eating hawk in the Everglades, and the effects that water management practices have had on environmental uncertainty (Beissinger 1986). Everglades water levels were found to be unpredictable for nesting Snail Kites (*Rostrhamus sociabilis*). When annual variation in water levels increased due to drainage and water management practices, predictability further declined. Colwell's index has also been used to relate the predictability of rainfall to habitat use by foraging herbivores (Low 1979) and human social systems (Low 1988), of temperature and precipitation to life history differences in ground squirrels (Zammuto and Millar 1985), and of streamflows as a metric to characterize flooding regimes (Gan et al. 1991).

In all of these applications, continuous data were categorized into environmental states to calculate Colwell's index. In such instances, the definition of environmental states can be problematic unless some biologically meaningful categories can be constructed. Both Colwell (1974) and Stearns (1981) discuss methods of categorizing continuous data to yield the most information about predictability. However, arbitrarily defining environmental states can cause problems because the number of states chosen influences the values of predictability, constancy, and contingency (Beissinger 1986, Gan et al. 1991). Also, the same patterns of precipitation or temperature could have different predictabilities depending on how organisms perceive and respond to fluctuations in their environment. To be valid, the states or categories chosen should be based on how the organism under consideration perceives the environment and the biological effects of those categories on the organism's life history. Because Colwell's index was not as sensitive as Fourier transformations in detecting periodicities with metric data (Stearns 1981), it may often be more valid to use time-series approaches which do not require categorization of environmental data to analyse periodicity.

Using spectral analysis to determine periodicity and seasonality

Time series techniques can be used with equispaced continuous data to examine environmental predictability by examining periodicity. Completely periodic fluctuations are highly predictable. Therefore, the strength of the periodicity of an environment is a measure of its predictability. Measures of periodicity should be equivalent to measures of contingency (M) in the sense of Colwell (1974).

Frequency domain or spectral approaches are the most useful for measuring periodicity. They break the time series into periodic cycles composed of regular sine and cosine waves. Spectral analysis assesses the fit of waves of different frequencies or bands to the time series data (Jenkins and Watts 1968, Gottman 1981). The spectral density function gives a measure of the intensity of the cycle at a given frequency for many frequencies in the time series. It can be thought of as the distribution of the variance of the time series that can be accounted for by cycles of different periods or lengths.

The spectra is estimated by calculating the autocovariance function, weighting with a lag "window," and then fitting the Fourier cosine transform (Platt and Denman 1975). The "window" improves the spectral density estimates by functioning like a weighted average and smooths the periodogram. The properties of many kinds of spectral windows (e.g., Bartlett, Parzen, Tukey-Hanning) have been investigated (Jenkins and Watts 1968, Anderson 1971, Gottman 1981), and most windows available on commercial computer packages are likely to be suitable.

Figure 1 depicts monthly rainfall for two time series and their corresponding spectral densities. The "random" rainfall pattern was created from a random numbers generator, while the "regular" time series was generated from a completely regular cycle with a 12 month period. The flat spectrum and small spectral density value for the random series are characteristic of "white noise" or unpredictable environments. The spectrum of the regular series shows a peak corresponding to a 12 month period which, due to its completely regular nature, occurs as a hump. Smaller peaks are due to shorter cycles (e.g., 2 months and 4 months) that are inherent in the annual cycle and are harmonics (Gottman 1981).

To measure the magnitude of seasonality of weather data, we are interested in evaluating the predictability of environmental conditions at the same time each year. Setting the spectral window to smooth a 12-month interval on either side of an observation affords the maximum opportunity to separate annual variation from other frequencies (Fig. 2). The 12-month spectral density estimate can then be used as a measure of seasonality or contingency. Because the magnitude of the spectral density is affected by the magnitude of the environmental fluctuations, the same time series measured in different units (e.g., inches, mm, cm) will give the same spectral pattern but different spectral densities. Thus, units must be standardized if spectral densities for time series of different environmental dimensions (e.g. temperature and rainfall) or locations are to be compared.

In addition to calculating the spectral density estimate for the period of 12 months, estimates will also be produced for a variety of other higher (< 12 months) and lower (> 12 month) frequencies (Fig. 2). This enables the detection of long-term cycles in the environment that may easily be overlooked. For example Beissinger (1986) showed how 5 to 7 year drought-flood cycles could be detected in a highly variable time series of 67 years of monthly Everglades water levels using spectral analysis. Incorporating these short- and long-term cycles into one measure of environmental predictability is beyond the reach of this paper. But low- frequency phenomena, such as catastrophes, are often of great interest despite their infrequent occurrence because they may be potent evolutionary forces. For example, changes in the annual cycle of Everglades water levels

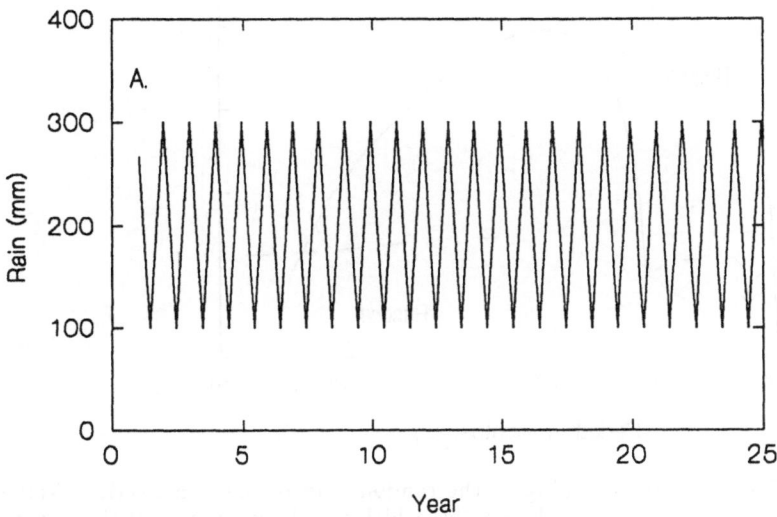

Figure 1A. A completely regular time series of monthly rainfall totals with a 12-month period.

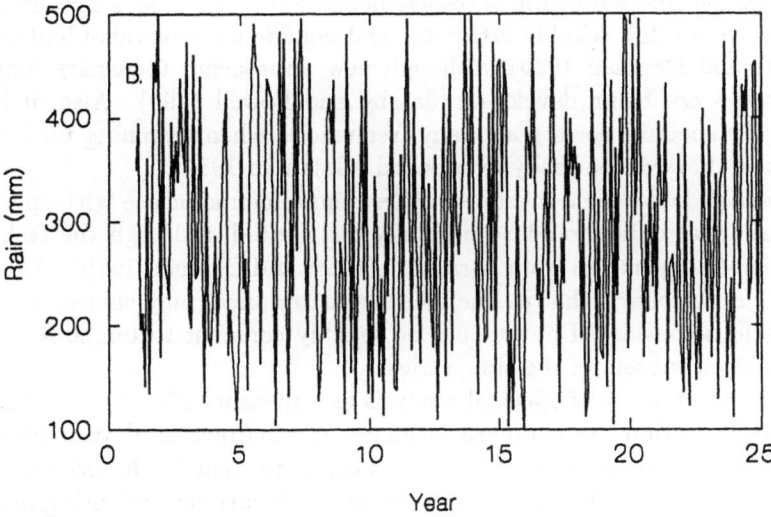

Figure 1B. A random time series of monthly rainfall totals created by a random numbers generator.

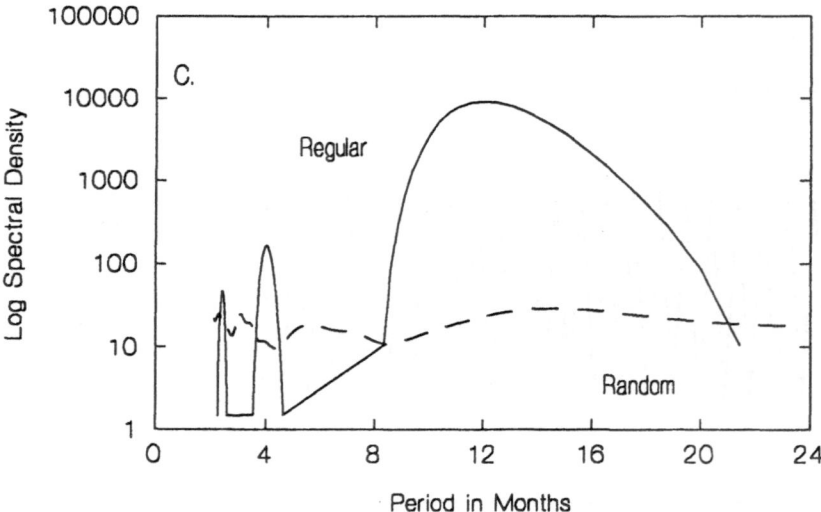

Figure 1C. Spectral densities corresponding to the random and regular time series. Values on the x axis are in months rather than cycle lengths, which are the reciprocal of the number of months in a cycle. Thus, the x axis has been reversed from typical plots of spectral analysis for ease of interpretation.

(a high- frequency phenomenon) have the most direct effect on Snail Kite nesting success and recruitment, whereas changes in the drought-flood cycle (a low-frequency phenomenon) may directly influence adult survival and life expectancy (Beissinger 1986). Low frequency cycles are also more difficult to resolve because spectral analysis requires 4-10 samples per cycle to develop reliable estimates and confidence intervals (Jenkins and Watts 1968, Platt and Denman 1975), although new approaches to resolve long cycles in short time series are being developed (Jassby and Powell 1990). Also, it is necessary to evaluate the spectral density at many frequencies when searching for low frequency cycles because they can be easily overlooked (Gottman 1981).

Finally, environmental data should meet two other requirements for use with spectral analysis. The time series data should be normally distributed, although the technique is robust to moderate departures from normality (Platt and Denman 1975). Also, the time series should be stationary, that is, the means and variances may change with the number of observations considered but should be roughly constant throughout the whole period for equal-sized subsets of the time series.

We have been exploring the use of spectral analysis as a measure of environmental predictabilty. In the next section, we compare estimates of environmental variability derived from Colwell's index and spectral analysis, and compare them both with standard statistical measures (means, estimates of variance, and autocorrelations) using real data sets. We expected to find a positive correlation between spectral density estimates of seasonality and Colwell's measure of contingency, and a negative correlation between spectral density estimates of seasonality and Colwell's measure of constancy. Measures of variation should be negatively correlated with constancy but positively related to contingency. Spectral densities might be expected to correlate positively with measures

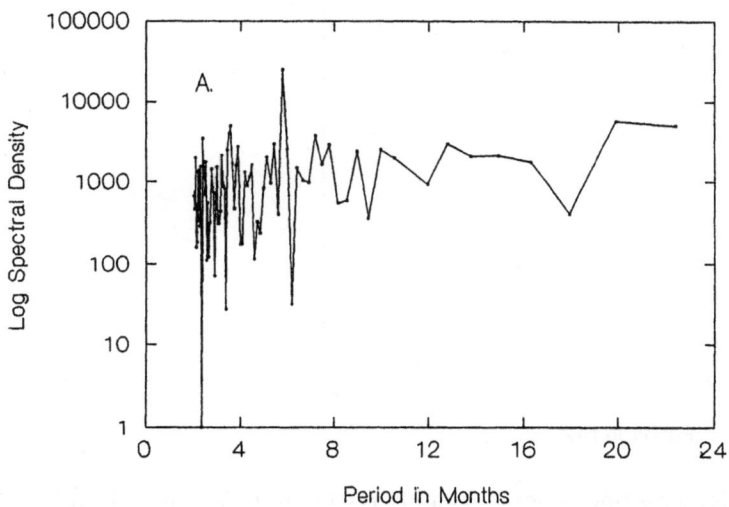

Figure 2A. Spectral densities for a time series of monthly rainfall records from Belem, Brazil without a smoothing window. *X*-axis as in Fig. 1C.

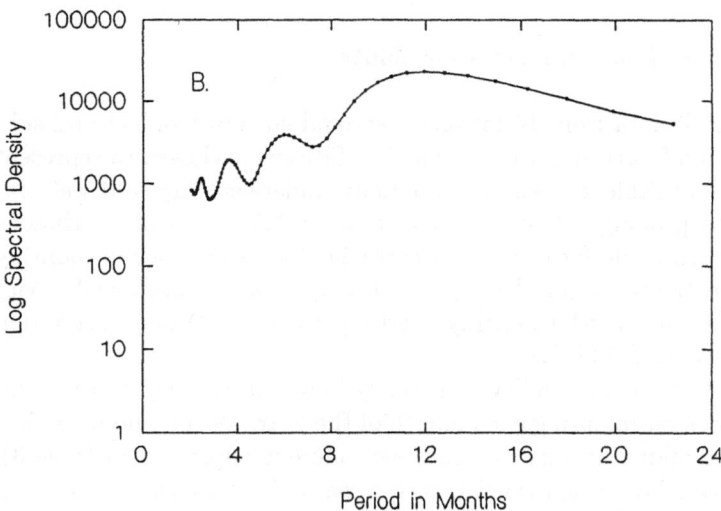

Figure 2B. Spectral densities for a time series of monthly rainfall records from Belem, Brazil, with a 12-month Parzen window to smooth the data and emphasize annual cycles.

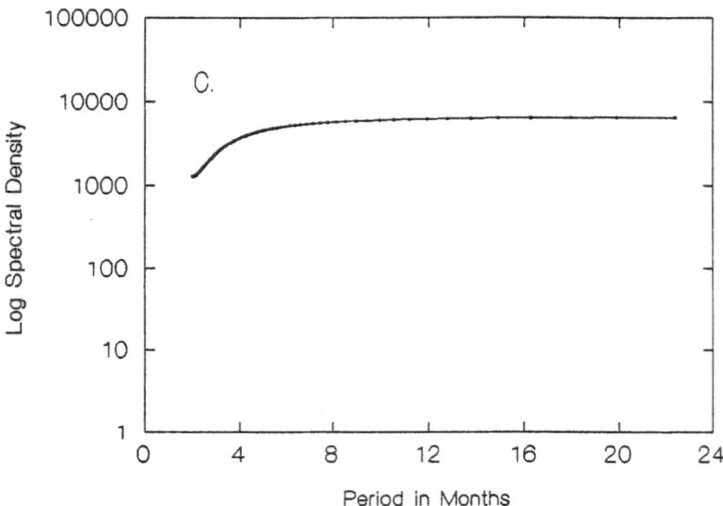

Figure 2C. Spectral densities for a time series of monthly rainfall records from Belem, Brazil with a 100-month Parzen window which precludes detection of periodicity.

of variation because seasonal environments are variable by nature, with means because spectral density increases with the magnitude of the units of the time series, and with autocorrelations because spectral densities are calculated using the autocovariance function.

Comparison among measures of environmental variability

We compiled monthly rainfall data from 15 lowland, wetland sites in tropical and subtropical portions of North and South America (Table 1). Sites were chosen to represent a variety of rainfall amounts (Table 2). All sites annually underwent dry and wet seasons, and none experienced prolonged frosts. Between 15 and 64 years of continuous monthly rainfall data were available for each site (Table 1). For each series of monthly rainfall data, we calculated the following descriptive statistics: mean, standard deviation, coefficient of variation, and lag-1 (monthly) and lag-12 (annual) autocorrelation coefficients of the rainfall values (Table 2).

Colwell's (1974) indices of predictability, constancy, and contingency were calculated using 12 time categories representing each month of the year. We had no biological criteria on which to base rainfall states so we used two different approaches (Table 3): scaled states and fixed states. For scaled rainfall states, we derived 7 classes of rainfall that were multiples of the mean ($\overline{X}$) of each series following the method of Gan et al. (1991): $\leq 0.5\overline{X}$, $0.5\overline{X}$–$1.0\overline{X}$, $1.0\overline{X}$–$1.5\overline{X}$, ..., $> 3.0\overline{X}$. For fixed rainfall states, we used the following absolute classes for classifying rainfall values: ≤ 10 mm, 11–50 mm, 51–100 mm, 101–250 mm, and ≥ 251 mm.

Table 1. Sites in tropical and subtropical North and South America included in our study of temporal patterns of monthly rainfall variation.

Site	Country	Latitude	Longitude
Buenos Aires	Argentina	34.6S	58.5W
Cuiaba	Brazil	15.6S	56.1W
Manaus	Brazil	3.1S	60.0W
Belem	Brazil	1.5S	48.5W
Central Farm	Belize	17.2N	89.0W
Belize Intl. Airport	Belize	17.5N	88.3W
Guayaquil	Ecuador	2.2S	79.8W
Corumba	Brazil	19.0S	57.7W
Asuncion	Paraguay	25.2S	57.5W
Puerto Casado	Paraguay	22.5S	57.9W
Tampico	Mexico	22.2N	97.9W
Veracruz	Mexico	19.2N	96.1W
Apure	Venezuela	7.9N	67.4W
Lake Okeechobee, Florida	U.S.A.	27.0N	80.7W
R3A, Florida	U.S.A.	25.8N	80.6W

Table 2. Characteristics of monthly rainfall patterns at the 15 study sites: mean monthly rainfall (mm), standard deviation of mean monthly rainfall (SD), coefficient of variation of monthly rainfall (CV), lag-1 and lag-12 autocorrelation coefficients ($r1$ & $r12$), and number of years of rainfall in the time series (Years).

Site	Mean	SD	CV	$r1$	$r12$	Years
Buenos Aires	86	61	71	0.020	0.008	39
Cuiaba	115	100	88	0.542	0.621	60
Manaus	171	113	66	0.582	0.619	53
Belem	245	152	62	0.602	0.612	15
Central Farm, Belize	123	97	78	0.308	0.378	22
Belize Intl. Airport	150	117	78	0.347	0.347	34
Guayaquil	78	130	165	0.567	0.584	31
Corumba	88	74	84	0.379	0.402	30
Asuncion	112	84	74	0.208	0.227	62
Puerto Casado	99	77	78	0.224	0.263	18
Tampico	86	125	144	0.168	0.427	20
Veracruz	146	193	132	0.464	0.580	20
Apure	118	128	108	0.618	0.664	64
Lake Okeechobee, FL	94	71	75	0.390	0.573	25
R3A, FL	131	93	71	0.236	0.443	25

Table 3. Characteristics of monthly rainfall patterns at the 15 study sites: Colwell's indices of predictability (P), constancy (C), and contingency (M), and $\log_{10}$-transformed spectral density for a period of 12 months (S). Colwell's indices were calculated based on scaled-rainfall states (e.g., P_s) and fixed-rainfall states (e.g., P_f).

Site	P_s	C_s	M_s	P_f	C_f	M_f	S
Buenos Aires	0.436	0.181	0.255	0.405	0.197	0.208	3.14
Cuiaba	0.650	0.148	0.502	0.715	0.069	0.646	4.61
Manaus	0.674	0.185	0.488	0.660	0.146	0.515	4.63
Belem	0.742	0.208	0.535	0.701	0.290	0.411	4.36
Central Farm, Belize	0.541	0.170	0.371	0.550	0.121	0.429	3.85
Belize Intl. Airport	0.555	0.167	0.388	0.579	0.126	0.454	4.24
Guayaquil	0.587	0.307	0.280	0.761	0.187	0.574	4.42
Corumba	0.568	0.156	0.413	0.587	0.092	0.496	3.91
Asuncion	0.472	0.166	0.306	0.454	0.148	0.306	3.91
Puerto Casado	0.523	0.174	0.349	0.522	0.160	0.361	3.50
Tampico	0.543	0.242	0.301	0.627	0.047	0.580	4.05
Veracruz	0.65	0.239	0.412	0.746	0.027	0.719	4.63
Apure	0.61	0.161	0.449	0.781	0.062	0.719	4.87
Lake Okeechobee, FL	0.618	0.171	0.447	0.591	0.158	0.433	3.78
R3A, FL	0.566	0.179	0.387	0.544	0.151	0.393	3.87

Spectral analysis was used to examine the relative strength of the annual rainfall cycle at each site. Spectral densities were estimated across a range of cycle periods using PROC SPECTRA of the SAS Institutes' collection of models for analyzing time series data (SAS Inst. 1984). A Parzen window with a truncation point of 12 was used to smooth the periodogram (Anderson 1971). The spectral density at 12 months was estimated and $\log_{10}$-transformed for comparisons with other indices of rainfall variation.

Relationships among indices of environmental variation were analysed using Spearman rank (r_s) correlation coefficients (Zar 1984). The effects of employing scaled- versus fixed-rainfall states on the values of Colwell's indices were assessed using the Wilcoxon signed-rank test on pairs of indices calculated for each site using the two methods.

As predicted, spectral density measures of seasonality were positively related to predictability and contingency estimates from both the scaled- and fixed-state calculations (Table 4). The expected negative correlation between spectral density and constancy was nearly significant for the fixed-state calculation but not for the scaled-state calculation. Significant correlations did not occur between the length of the time series and any of the measures of environmental variability.

Comparisons among Colwell's indices revealed that predictability and contingency were strongly and positively correlated at the sites that we analyzed (Table 4). This pattern held for both scaled-state and fixed-state calculations. For the scaled-state measures, neither constancy and predictability nor constancy and contingency were correlated with one another, whereas for the fixed-state measures constancy and contingency

Table 4. Spearman rank correlations among monthly rainfall statistics. See tables 2 & 3 for definitions of the abbreviations used in row and column headings. Spearman correlations ≥ 0.518 are statistically significant ($P \leq 0.05$) and are presented in bolded text.

	Mean	SD	CV	r1	r12	P_s	C_s	M_s	P_f	C_f	M_f	S
Mean	1.000											
SD	0.422	1.000										
CV	-0.469	0.386	1.000									
r1	0.438	**0.589**	0.090	1.000								
r12	0.325	**0.604**	0.203	**0.886**	1.000							
P_s	0.497	**0.552**	-0.014	**0.878**	**0.883**	1.000						
C_s	-0.061	0.439	0.041	-0.021	0.064	0.129	1.000					
M_s	**0.647**	0.264	-0.237	**0.757**	**0.711**	**0.835**	-0.343	1.000				
P_f	0.154	**0.768**	**0.522**	**0.843**	**0.896**	**0.788**	0.186	0.511	1.000			
C_f	-0.100	-0.314	**-0.603**	-0.086	-0.261	-0.138	0.275	-0.196	-0.379	1.000		
M_f	0.064	**0.645**	**0.715**	**0.599**	**0.731**	**0.593**	0.032	0.388	**0.879**	**-0.736**	1.000	
S	0.409	**0.796**	0.357	**0.786**	**0.796**	**0.706**	0.046	**0.539**	**0.861**	-0.500	**0.822**	1.000

were strongly and negatively correlated while constancy and predictability remained unrelated.

Calculating Colwell's indices based on scaled- versus fixed-rainfall states had no effect on the magnitude of the predictability index (Wilcoxon signed ranks test, $z = 1.079$, $P = 0.281$, $n = 15$), but shifted the relative contributions of constancy and contingency to predictability. Constancy values decreased ($z = -2.613$, $P = 0.009$) and contingency values increased ($z = 2.229$, $P = 0.026$) when the indices were calculated using the fixed-state method in comparison to calculations based on the scaled-state method.

Standard statistical measures were equivocal descriptors of environmental predictability (Table 4). Average annual rainfall was unrelated to most measures of predictability. The standard deviation of the rainfall series was positively correlated with spectral densities, with both calculations of Colwell's index of predictability, and with the fixed-state calculation of contingency. But, the standard deviation was not significantly negatively related to either calculation of constancy. The coefficient of variation was unrelated to spectral densities and scaled-state Colwell estimates, but was correlated positively with contingency and negatively with fixed-state Colwell estimates. Spectral density and contingency estimates from both scaled- and fixed-state calculations were correlated positively with the lag-1 and lag-12 autocorrelation coefficients, but there was no relationship between autocorrelation coefficients and constancy.

Discussion of measures of environmental predictability

Two measures of environmental predictability derived in very different manners (e.g., from information theory indices and spectral analysis) were relatively concordant with one another for the 15 series of rainfall that we analyzed. Notably, estimates of seasonality and predictability based on spectral densities were correlated positively with estimates derived from Colwell's index of predictability and contingency, and to some extent negatively with constancy. Stearns (1981) also found similar correlations between measures derived from Fourier analyses, and constancy and contingency.

Colwell's index has the advantage of providing an absolute scale for environmental predictability compared to the relative scale of spectral densities values. Furthermore, unlike spectral analysis, Colwell's index can discern the contributions of contingency and constancy to predictability. The decomposition of predictability into different components could be a useful approach to study how selection acts on of the evolution of characters in constant, seasonal and stochastic environments. But results from Colwell's index are more sensitive than spectral analysis to data handling decisions made by the researcher and to the problems of categorization of metric data (Stearns 1981). This was illustrated in our analysis by differences in constancy and contingency estimates using environmental states based on fixed categories versus those calculated using a sliding scale around the mean (Gan et al. 1991). A sliding scale is useful when there are no clear categories for environmental states, but it tends to increase estimates of constancy and decrease seasonality (contingency). Since the significance of environmental predictability or stochasticity is relative to other environments, it is probably more meaningful to keep environmental states constant across different sites (i.e., fixed-state calculations).

Because many environmental characteristics are measured by metric data, spectral analysis may offer the best choice for characterizing predictability in the absence of any clear criteria for categorizing environmental states. Furthermore, spectral analysis permits assessment of other cyclical patterns of variation in the time series that may have important life history implications. Further exploration of the behavior of spectral density as an estimator of predictability is needed.

In the particular environments that we analysed, the relative inputs of Colwell's constancy and contingency indices suggested that the index of predictability was strongly dependent on the seasonality of the rainfall. Two factors, one mathematical and the other biological, can account for the high positive correlation in this study between Colwell's predictability and contingency indices. First, predictability is the sum of constancy and contingency (Colwell 1974). Second, we focused our analyses on environments where rainfall varies among months (i.e., is not particularly constant) but occurs predictably in certain months (i.e., is contingent on a particular month of the year). Thus, predictability was necessarily more a function of contingency than constancy for the locations that we analyzed. In contrast, analysis of other environments indicated that constancy (analysis of data in Low 1979), or both constancy and contingency (Gan et al. 1991) contribute significantly to environmental predictability.

Attempts to describe environmental variation and understand its effects on life-history evolution have often employed simplistic measures of variability, such as standard deviations or coefficients of variation. In our analysis of relatively seasonal environments (Table 4), some standard statistical measures (e.g., standard deviation and autocorrelation coefficients) were sometimes correlated with time series measures of predictability (e.g., Colwell's contingency or spectral densities) but others were not (e.g., mean and coefficient of variation). Some degree of correlation between measures of variation and predictability should be expected in environments where predictability arises mostly from contingency. However, standard statistical measures have been shown to be completely inadequate as measures of environmental predictability in environments where predictability arises mostly from constancy or the combination of constancy and contingency (Low 1979, Gan et al. 1991). Thus, we implore ecologists and evolutionary biologists to employ times series methods that are more sophisticated than simple statistical measures if their studies seek to characterize environmental variation and explore its implications for life-history evolution.

References

Anderson, T.W. 1971. *The Statistical Analysis of Time Series*. Wiley, New York.

Beissinger, S.R. 1986. Demography, environmental uncertainty, and the evolution of mate desertion in the Snail Kite. *Ecology* 67: 1445–1459.

Colwell, R.K. 1974. Predictability, constancy, and contingency. *Ecology* 55: 1148–1153.

Eckert, K.L. 1987. Environmental unpredictability and leatherback sea turtle (*Dermochelys coriacea*) nest loss. *Herpetologica* 43: 315–323.

Futuyma, D.J. 1986. *Evolutionary Biology*. 2nd edition. Sinauer, Sunderland, Ma.

Gan, K.C., T.A. McMahon, and B.L. Finlayson. 1991. Analysis of periodicity in streamflow and rainfall data by Colwell's indices. *J. Hydrology* 123: 105–118.

Gottman, J.M. 1981. *Time-Series Analysis*. Cambridge University Press, Cambridge.

Jasby, A.D., and T.M. Powell. 1990. Detecting changes in ecological time series. *Ecology* 71: 2044–2052.

Jenkins, G.M., and D.G. Watts. 1968. *Spectral Analysis and its Applications*. Holden-Day, San Francisco.

Low, B.S. 1978. Environmental uncertainty and the parental strategies of marsupials and placentals. *Amer. Nat.* 112: 197–213.

Low, B.S. 1979. The predictability of rain and the foraging patterns of the red kangaroo (*Megaleia rufa*) in central Australia. *J. Arid Envir.* 2: 61–76.

Low, B.S. 1988. Human responses to environmental extremeness and uncertainty: a cross-cultural perspective. In E. Cashdan (Ed.) *Risk and Uncertainty in Tribal Peasant Economies*. Westview Press, Boulder. Pp. 229–255.

Newton, I. 1990. *Lifetime Reproduction in Birds*. Academic Press, London.

Palumbi, S.R. 1984. Tactics of acclimation: morphological changes of sponges in an unpredictable environment. *Science* 225: 1478–1480.

Slater, R.O. 1962. Climate of the Alice Springs area. In R.A. Perry (Ed.) *Lands of the Alice Spring area, Northern Territory*. CSIRO Land Research. Ser. 6., Melbourne. Pp. 109–128.

Sokal, R.R., and F.J. Rolf. 1981 *Biometry*. 2nd edition. W.H. Freeman, New York.

Stearns, S.C. 1981. On measuring fluctuating environments: predictability, constancy, and contingency. *Ecology* 62: 185–199.

Wrege, P.H., and S.T. Emlen. 1991. Breeding seasonality and reproductive success of White-fronted Bee-eaters in Kenya. *Auk* 108: 673–687.

Zack, S., and K.N. Rabenold. 1989. Assessment, age and proximity in dispersal contests among cooperative wrens: field experiments. *Anim. Behav.* 38: 235–247.

Zammuto, R.M., and J.S. Millar. 1985. Environmental predictability, variability, and *Spermophilus columbianus* life history over an elevational gradient. *Ecology* 66: 1784–1794.

Zar, J.H. 1984. *Biostatistical Analysis*. 2nd edition. Prentice-Hall, Englewood Cliffs, N.J.

MODELING SELECTION ON CONDITIONAL STRATEGIES

IN STOCHASTIC ENVIRONMENTS

Wade N. Hazel
Department of Biological Sciences

and

Richard Smock
Department of Mathematics and Computer Science
DePauw University
Greencastle, IN 46135

Abstract. Conditional strategies, in which individuals can exercise different tactics depending on the circumstances, are believed to be adaptive when the environment varies stochastically. We discuss a genetic model that treats such strategies as threshold traits. Examples are provided that illustrate how the model can be used to examine the effects of selection on conditional strategies when the environment varies spatially. The implications of the model for maintenance of conditional strategies in stochastic environments are discussed.

1. Conditional strategies

The ability to look, function or behave in distinctly different ways depending on the situation is one way in which organisms have adapted in environments that vary stochastically. In the terminology of evolutionary game theory this ability is defined as a conditional strategy (Dawkins 1980).

Conditional strategies consist of two or more tactics that individuals can exercise depending on the situation. Examples can be drawn from all areas of biology and include predator-induced polymorphisms common in a wide variety of aquatic invertebrates (reviewed in Dodson 1989, Harvell 1990), environmentally-cued (developmental) polymorphisms in color and/or morphology in insects (Smith 1978, Hazel and West 1979, Sims 1983, Greene 1988), and size dependent reproductive strategies common in both vertebrates and invertebrates (Krebs and Davies 1987). Some examples represent adaptations to environments that vary stochastically in space, while others are adaptive when conditions vary temporally.

Theoretical studies of the evolution and maintenance of conditional strategies have typically relied on phenotypic or strategy models (for examples see Lloyd, 1984; Lively 1986c; Moran 1992). The simplest specify a distribution of environments in which individuals find themselves (e.g. coarse-grained or fine-grained, temporally variable or spatially variable), a distribution of possible strategies (e.g. two or more unconditional strategies and one or more conditional strategies), and fitnesses for these strategies that vary depending on the environment. In addition, we might assume that a conditional strategy is intrinsically more costly than non-conditional strategies. Or, we might wish to vary the accuracy of the environmental cue on which the conditional strategy is dependent. Once the assumptions are specified one would then determine, either analytically or using simulation, under what circumstances a conditional strategy can increase in frequency when rare and resist invasion by other strategies when common (i.e. become an ESS).

Overall, the results of such models appear to be in close agreement with what is seen in nature. Thus, the picture that emerges from the combined results of theoretical (Levins 1963; Lloyd 1984; Lively 1986c; Moran 1992) and empirical studies (West and Hazel 1979; Lively 1986a,b) is that conditional strategies are favored by selection when 1) environmental variation is coarse grained (spatially or temporally), 2) there is a fitness tradeoff for the alternative tactics in different environments, and 3) reliable cues are associated with the environmental variation. However, what such studies have generally failed to address is how, once evolved, genetic variation underlying a conditional strategy will be affected by selection in a stochastic environment.

The purpose of this contribution is to describe a genetic model that begins with the realistic assumption of genetic variation underlying conditional strategies. Given this assumption, we then illustrate how the model allows one to analytically determine, using conventional techniques of quantitative genetics, how conditional strategies will respond to varying selection in a stochastic environment. Finally, we discuss the general implications of the model for the study of conditional strategies in stochastic environments.

2. Conditional strategies as threshold traits

The genetic model most applicable to conditional strategies was originally developed by Wright (1920) to explain all-or-none phenotypic variation that is not inherited as a simple Mendelian trait. The basic model, described in detail in Falconer (1981) assumes that the expression of an all-or-none character is dependent on some underlying continuous phenotypic variable. If this variable exceeds some threshold value then one phenotype is expressed, if it fails to exceed the threshold value, then another phenotype is expressed; hence the name threshold trait.

To adapt the threshold model to conditional strategies we have generalized the work of Bulmer and Bull (1982) on environmental sex determination. The major differences between our model (Hazel et al. 1990) and the description in the preceding paragraph is that we assume that the threshold value is a function of the environment, which is assumed to vary stochastically and independently of the hypothesized phenotypic variable, which we define as the reaction norm switch point. Thus, the threshold values, which can vary continuously or discontinuously depending on whether the environment varies as discrete patches or continuously, truncate the continuous distribution of the phenotypic variable into discrete character states representing the tactics comprising the conditional strategy.

For example, consider a population in which the underlying phenotypic variable is a random variable X with expected value μ and heritability h^2. According to our model, individuals would exercise tactics A or B in response to some environmental cue which is a random variable T, such that individuals exercise tactic A if $x > t$ and tactic B if $x < t$. Therefore, genetic variation in the phenotypic variable, X, is equivalent to genetic variation underlying the conditional strategy. Selection on such variation occurs according to some fitness criterion with the result that the underlying phenotypic distribution in the reproducing population will be different from that of the general population. If we denote the expected value of X in the nth generation by μ_n then the expected value of the underlying phenotypic distribution of generation $n + 1$ is given by

$$\mu_{n+1} = \mu_n + h^2 S(\mu_n)$$

where $S(\mu)$ is the selection differential and is equal to the expected value of X after selection minus the expected value of X before selection. To understand the nature of $S(\mu)$ let $f(x, \mu)$ be the probability density function for X (we will need to exploit the fact that f depends upon μ), let $g(t)$ be the probability density function for T and let $w(x, t)$ denote the fitness function. Assuming X and T are independent, the joint probability density function of X and T before selection is $f(x, \mu)g(t)$ and the joint density function after selection is given by $\frac{f(x,\mu)g(t)w(x,t)}{D(\mu)}$ where

$$D(\mu) = \int_{-\infty}^{\infty} f(x, \mu) \int_{-\infty}^{\infty} g(t)w(x, t) \, dt \, dx. \tag{1}$$

Note that D is a normalization factor which reflects the effect of the fitness function on the joint probability density function. Since the expected value of X before selection is μ_n and the expected value of X after selection is

$$\frac{\int_{-\infty}^{\infty} xf(x, \mu_n) \int_{-\infty}^{\infty} g(t)w(x, t) \, dt \, dx}{D(\mu_n)}$$

we have

$$S(\mu) = \frac{C(\mu)}{D(\mu)}$$

where

$$C(\mu) = \int_{-\infty}^{\infty} (x - \mu)f(x, \mu) \int_{-\infty}^{\infty} g(t)w(x, t) \, dt \, dx. \tag{2}$$

3. Applications

Discrete environments. Suppose the environment consists of two patches which we refer to as environments 1 and 2. Then the environmental variable T takes discrete values t_1 or t_2 with probabilities q and $1 - q$ respectively, depending on which patch is entered. Selection acts through the fitness function

$$w(x,t) = \begin{cases} \alpha_1, & t = t_1 \text{ and } x > t_1 \\ \alpha_2, & t = t_2 \text{ and } x > t_2 \\ \beta_1, & t = t_1 \text{ and } x < t_1 \\ \beta_2, & t = t_2 \text{ and } x < t_2 \end{cases}.$$

We can represent the probability density function of T as $g(t) = q\delta(t - t_1) + (1 - q)\delta(t - t_2)$ where $\delta(t)$ is is the Dirac delta function and is defined by its action in an integral:

If $\phi(t)$ is a function which is bounded and integrable on $(-\infty, \infty)$ and continuous at $t = \tau$ then

$$\int_{-\infty}^{\infty} \phi(t)\delta(t - \tau)\, dt = \phi(\tau).$$

If we assume that X is normally distributed with variance σ_x^2, then substituting the above expressions for g and w into (1) and (2) yields

$$S(\mu) = \frac{\sigma_x^2}{D(\mu)}[(\alpha_1 - \beta_1)qf(t_1,\mu) + (\alpha_2 - \beta_2)(1-q)f(t_2,\mu)] \tag{3}$$

where now

$$D(\mu) = \alpha_1 q p_1 + \beta_1 q(1 - p_1) + \alpha_2(1-q)p_2 + \beta_2(1-q)(1-p_2)$$

and $p_i = \int_{t_i}^{\infty} f(x,\mu)\, dx$. Expression (3) can be used to determine the equilibrium mean of the underlying phenotypic variable, x. To do this, one needs estimates of the fitnesses of the alternative tactics in each environment, the frequencies with which individuals enter the environments, and the variance of the underlying phenotypic variable. For example, if the value of the cue can be modified experimentally, then the cumulative frequencies with which the alternative tactics are exercised for different values of the cue can be used to estimate σ_x^2. By substitution into expression (3), these estimates can be used to determine the value of μ for which $S(\mu) = 0$:

$$\mu = \frac{t_1 + t_2}{2} + \frac{\sigma_x^2}{t_2 - t_1} \ln \frac{q(\alpha_1 - \beta_1)}{(1 - q)(\beta_2 - \alpha_2)}.$$

This value of μ is the equilibrium or steady state mean of the distribution of the phenotypic variable.

The first term in this expression is the arithmetic mean of the two environments; the second term is a measure of the displacement from this arithmetic mean. If $q(\alpha_1 - \beta_1) =$

$(1-q)(\beta_2 - \alpha_2)$ (i.e. the intensity of selection is the same in the two environments), then the equilibrium mean is the arithmetic mean. If selection is more intense in environment 1 ($|q(\alpha_1 - \beta_1)| > |(1-q)(\beta_2 - \alpha_2)|$) then the equilibrium mean shifts in the direction of t_2, and this results in a greater proportion of the population exercising the appropriate tactic in environment 1 than in environment 2. The reverse occurs if selection is more intense in environment 2.

The equilibrium mean is stable provided tactic A is better than tactic B in environment 1 ($\alpha_1 > \beta_1$). This implies that B is better than A in environment 2 ($\beta_2 > \alpha_2$); otherwise $S(\mu) = 0$ has no solution. The stability result is obtained by observing that $\frac{dS}{d\mu} < 0$ at the equilibrium.

The theoretical equilibrium can be compared to the observed mean since the latter can be estimated from the cumulative frequency distribution as the cue value for which a 1:1 ratio of the alternative tactics exists. Close agreement between the theoretical and observed mean would suggest that the conditional strategy is being maintained at equilibrium by selection.

The model can be used even if it is not possible to readily quantify the environmental cue and thus generate a cumulative frequency distribution for the phenotypic variable. Needed are estimates of the fitnesses of the alternative tactics in each environment, the frequencies with which individuals enter the environments, and the frequencies with which individuals exercise the alternative tactics in each environment. The frequencies with which the alternative tactics are exercised in the environments can be used to quantify (in standard deviation units) both t_1 and t_2 relative to the mean phenotypic value, μ. For example, if in environment 1, 94% of individuals exercise tactic A, then t_1 must lie 1.56 standard deviation units from μ. If in environment 2, 6% of individuals exercise tactic A, then t_2 must lie 1.56 standard deviation units to the other side of μ. Thus, t_1 and t_2 must lie 3.12 standard deviation units apart, and μ must lie halfway between t_1 and t_2. Based on these estimates, we can arbitrarily set our observed values for t_1, t_2, and μ at 1.00, 4.12 and 2.56 standard deviation units. Substituting our estimates for t_1 and t_2, the fitnesses of the alternative tactics in each environment, the frequencies with which individuals enter the environments into expression (3) and solving for μ at which $S(\mu) = 0$ gives the predicted equilibrium value of μ (in standard deviation units). Agreement between observed and predicted values suggests that the conditional strategy is being maintained at equilibrium by selection.

Continuous environments. If the environment varies continuously, then the environmental variable T is continuous, and selection operates through the fitness function

$$w(x,t) = \begin{cases} W_A(t), & x > t \\ W_B(t), & x < t \end{cases}.$$

Under the assumption that the phenotypic variable X is normally distributed we can find the equilibrium mean of X by solving

$$\int_{-\infty}^{\infty} g(t)f(t,\mu)W_A(t)\,dt = \int_{-\infty}^{\infty} g(t)f(t,\mu)W_B(t)\,dt \tag{4}$$

for μ.

This application of the model can be readily used to investigate the effects of selection on size-dependent conditional male reproductive tactics. For example, consider those conditional strategies where, depending on their size, males exercise different tactics (e.g fighting or sneaking) to gain access to females (see Alcock et al. 1977, Gross 1985 for examples). For such traits, body size can be treated as the environmental variable, T. Field observations can be used to estimate the fitness functions of the alternative tactics, $W_A(t)$ and $W_B(t)$, and the probability density function, $g(t)$, of male body size. The cumulative frequencies with which the alternative tactics are exercised in males of different body sizes can be used to estimate the variance of the phenotypic variable. Substituting these estimates into expression (4) and solving for μ gives the predicted mean of the phenotypic variable in the units used to quantify body size. Agreement between this value and the observed value for μ (estimated by the body size at which a 1:1 ratio of alternative tactics obtains) would be evidence that the conditional strategy is being maintained at equilibrium by selection. (See Hazel et al. 1990 for an example of this application.)

4. Implications and extension

The general implications of our model for the conditions necessary for the evolution of conditional strategies are consistent with those of phenotypic models. Both suggest that when perceptible environmental cues are correlated with coarse-grained environmental variation, and when fitness varies with the environment, selection will favor the evolution and maintenance of a conditional strategy. However, unlike phenotypic models, our model makes quantitative predictions for how a conditional strategy will respond to selection in a stochastic environment. That is, given estimates of the variation in the underlying phenotypic variable, the form and amount of stochastic variation in the environment and the fitnesses of the alternative tactics across those environments, our model can predict the equilibrium distribution of the phenotypic variable underlying the conditional strategy.

Implicit in our model is that, given the conditions outlined above, at equilibrium the conditional strategy will be maintained by stabilizing selection. Surprisingly, even a small amount of heterogeneity can still result in the maintenance of the conditional strategy. For example, consider the hypothetical situation where the phenotypic variable underlying the conditional strategy is initially distributed with $\mu = 0$ and $\sigma_x^2 = 1$. Assume environments are discrete and t_1 and t_2 are +2 and +6 standard deviations from μ, and that the probability of entering environment 1 is 0.99. If the fitness differences between the two tactics in the two patches are equivalent then, using expression 3, selection on the phenotypic variable will lead to an equilibrium mean of $\mu = 5.15$. Thus, even though the probability of entering environment 2 is small, approximately 80% of the population would have phenotypic values that lie between t_1 and t_2, and therefore be capable of exercising the more adaptive of the alternative tactics in each environment.

Although we have emphasized how the model can be applied to selection on conditional strategies in a spatially variable environment, it should be equally applicable to conditional strategies that are adaptations to temporal variation (e.g. seasonal polyphenisms). For example, consider the conditional strategy studied by Hairston and his colleagues involving the production of diapausing versus subitaneous eggs by the copepod *Diaptomus sanguineus* (see Hairston and Dillon 1990, and references therein). In response to photoperiod, females switch from producing subitaneous eggs to producing diapausing eggs in the spring prior to the onset of seasonal predation on the eggs by fish. However, the timing of the onset of predation varies somewhat from year to year, and hence the date (photoperiod) at which it is most adaptive to begin producing diapausing eggs varies as well. Variation in the response to photoperiod is heritable and is analogous to the phenotypic variable in our model. The long term effects of natural selection on this variable will depend on the historical pattern of the timing of the onset of seasonal predation relative to the life cycle of the animal and the relative costs of producing diapausing eggs early or late. Although this application is beyond the scope of this paper, it should be possible, given estimates of the above variables, to predict the equilibrium mean of phenotypic variable underlying the response to photoperiod.

We believe our model is potentially applicable to any situation in which organisms make "decisions" in response to environmental cues, even when those cues are other aspects of an organism's own phenotype. For example, in applying the model to size-dependent male reproductive tactics (see Hazel et al. 1990), we have treated male body size as the environmental variable. Such as application is valid because from the point of view of the genes underlying size-dependent male reproductive strategies the size of the bodies in which they find themselves is just another stochastic environmental variable.

Acknowledgements

This research was supported by a grant from the Howard Hughes Medical Institute to DePauw University. The comments of two external reviewers improved this manuscript. We thank them all.

References

Alcock, J., Jones, C.E. & Buchmann, S. L. 1977. Male nesting strategies in the bee *Centris pallida* Fox (Anthophoridae: Hymenoptera). *Am. Nat.* 111:145–155.

Bulmer, M.G. & Bull, J.J. 1982. Models of polygenic sex determination and sex ratio control. *Evolution* 36:13–26.

Dawkins, R. 1980. Good strategy or evolutionary stable strategy? In *Sociobiology: Beyond Nature/Nurture* (eds. Barlow, G.W. & Silverberg, J.), pp. 331–367. Boulder, Colorado: Westview.

Dodson, S.I. 1989. Predator induced reaction norms. *Bioscience.* 39:447–452.

Falconer, D.S. 1981. *Introduction to Quantitative Genetics*. London, England: Longman.

Greene, E. 1988. A diet-induced developmental polymorphism in a caterpillar. *Science* 243:643–646.

Gross, M. 1985. Disruptive selection for alternative life histories in salmon. *Nature* 313:47–48.

Harvell, C.D. 1990 The ecology and evolution of inducible defences. *Quarterly Review of Biology* 65:323–340.

Hairston, N.G., Jr. & Dillon, T.A. 1990. Fluctuating selection and response in a population of freshwater copepods. *Evolution* 44:1796–1805.

Hazel, W.N. & West, D.A. 1979. Environmental control of pupal colour in swallowtail butterflies (Lepidoptera: Papilioninae). *Ecological Entomology* 4:393–400.

Hazel, W.N., Smock, R., and Johnson, M.D. 1990. A polygenic model of the evolution and maintenance of conditional strategies. *Proc. Roy. Soc. Lond.* B 242:181–187.

Krebs, J.R. & Davies, N.B. 1987. *An Introduction to Behavioural Ecology.* Oxford, England: Blackwell.

Levins, R. 1963. Theory of fitness in a heterogeneous environment. II. Developmental flexibility and niche selection. *Am. Nat.* 97:75–90.

Lively, C.M. 1986a. Predator-induced shell dimorphism in the acorn barnacle *Chthamalus Canisopoma. Evolution* 4:232–242.

Lively. C.M. 1986b. Competition, comparative life histories, and maintenance of shell dimorphism in a barnacle. *Ecology* 67:858–864.

Lively, C. M. 1986c. Canalization versus developmental conversion in spatially variable environment. *Am. Nat.* 128:561–572.

Lloyd, D.G. 1984. Variation strategies of plants in heterogeneous environments. *Biol. J. Linn. Soc.* 21:357–385.

Moran, N.A. 1992. The evolutionary maintenance of alternative phenotypes. *Am. Nat.* 139:971–989.

Sims, S.R. 1983. The genetic and environmental basis of pupal colour dimorphism in *Papilio zelicaon. Heredity* 5:159–168.

Smith, A.G. 1978. Environmental factors influencing pupal colour in Lepidoptera. I. Experiments with *Papilio polytes, Papilio demoleus,* and *Papilio polyxenes. Proc. Roy. Soc. Lond., B.* 2:295–329.

West, D.A. & Hazel, W.N. 1979. Natural pupation sites of swallowtail butterflies (Lepidoptera: Papilioninae): *Papilio Apolyxenes* Fabr., *P. glaucus* L., *Battus philenor* (L.). *Ecological Entomology* 4:387–392.

Wright, S. 1920. The relative importance of heredity and environment in determining the piebald pattern of guinea pigs. *Proc. Nat. Acad. Sci.* 6:320–332.

COEXISTENCE IN STOCHASTIC ENVIRONMENTS

THROUGH A LIFE HISTORY TRADE OFF IN DROSOPHILA

Jan G. Sevenster and Jacques J.M. van Alphen
Populatiebilogie, Zoölogisch Laboratorium
Universiteit Leiden
Postbus 9516, NL 2300 RA Leiden, The Netherlands

Summary. The coexistence of competing species may be mediated by various mechanisms including resource partitioning and various kinds of environmental heterogeneity. In this paper we show how differences in life history enhance coexistence in stochastic environments. In *Drosophila* species, as in many other taxa, the developmental period is proportional to adult survival. A short developmental period, i.e. a high developmental rate, enhances the competitive ability of larvae. High adult survival, on the other hand, must increase the probability of reaching new breeding sites in space and time. We present a model of two competing species to investigate the consequences of the trade off. The model features density dependent mortality (due to competition) in the larval stage, and age dependent mortality in the adult stage. Breeding opportunities occur with a certain probability per time step. This is the only stochastic component of the model. The model demonstrates that fast growing, short lived species are superior when breeding opportunities are frequent. Slower growing, long lived species are superior when breeding opportunities are rare in time. A sensitivity analysis indicates that this conclusion is qualitatively robust. The mutual invasibility criterion reveals that stable coexistence will occur for certain feeding probabilities.

Introduction

The coexistence of competitors is an important issue in ecology. According to the principle of competitive exclusion (Hardin 1960), species using the same resources can not coexist. Nevertheless, competition does not always have a clear influence on the composition of communities (Strong et al. 1984). Models show that aggregation over patches (Shorrocks et al. 1984), environmentally induced variation in competitive ability (Begon and Wall 1987), as well as spatial and temporal heterogeneity in the environment (e.g. Chesson 1985a, Comins and Noble 1985) may lead to coexistence of species that share resources. In this paper, we will show how differences in life history provide a mechanism enhancing coexistence of species using the same resources in a temporally variable environment.

Within taxa at the class or family level, the ratio of the developmental period to the adult life span appears to be constant (Charnov and Berrigan 1990). The underlying reason may be a common dependence of many life history traits on metabolic rate and/or body size (Atkinson 1979, Hoffmann and Parsons 1989, Treveleyan et al. 1990, Harvey et al. 1991). Here we will investigate the consequences of the trade off between developmental rate and adult survival for coexistence of *Drosophila* species. *Drosophila* oviposit on ephemeral patches of food (e.g. decaying or fermenting fruits). Larvae stay on the patch and may have to compete for food. After pupation, adult flies emerge and have to find new breeding sites.

In *Drosophila*, a high developmental rate enhances larval competitive ability, because faster developing larvae are more likely to complete development before the patch is exhausted (e.g. Bakker 1961). However, fast growing species will produce short lived adults, that may not live to find a new breeding site in times of scarcity. These "fast" species will be superior when larval competition is strong and breeding sites are easily found. Although slowly growing species will be bad larval competitors, their higher adult survival will improve the chances of reaching new breeding sites in space and time. These "slow" species will therefore be superior when breeding sites are scarce and larval competition is weak. Such changes in superiority may enhance the coexistence of species using the same resources in variable environments (Chesson 1985a,b).

In this paper, we model the population dynamics of two competing species with different strategies in a stochastically varying environment. We test if superiority does indeed shift from fast to slow species when food becomes scarcer, and whether this could lead to stable coexistence. The hypotheses and the model are motivated and supported by field observations and experimental data of a guild of frugivorous *Drosophila* from a tropical rainforest in Panama (Sevenster and Van Alphen, *in prep.*). In this group of species, opposite strategies are found in an environment where the abundance of fruits varies greatly (Foster 1982, Sevenster and Van Alphen *in prep.*).

Model

The present simulation model describes competition between a fast and a slow *Drosophila* species when larval food is intermittently available. It hinges on two phenomena. First, the developmental rate determines larval competitive ability (Bakker 1961, 1969; models by De Jong 1976, Nunney 1983). Second, the inverse of the developmental rate, the developmental period α, is proportional to the adult life span λ (Atkinson 1979; Charnov and Berrigan 1990; Sevenster and Van Alphen *in prep.*).

The model has discrete time steps. Reproductive opportunities (i.e. the presence of fallen fruits) arise once every i time ps (days) on average. The adults of both species then produce offspring. The outcome of competition between the offspring is modelled, and survivors are added to the adult population after their developmental period. The adults are kept in one-day age classes and they experience age specific survival. Density dependent mortality thus occurs in the larval stage, and density independence in the adult stage. We will now describe the model in more detail.

Adult survival is age dependent and decreases with increasing age. Our experimental data on adult survival (Sevenster and Van Alphen, *in prep.*) are described best with the Erlang distribution. This distribution is known as the multiple hit model. It presumes that an organism accumulates randomly occurring deleterious events, or "hits," until it reaches a threshold number n and dies. The distribution has the density function (Richter and Söndgerath 1990)

$$f(t) = \beta \frac{(\beta t)^{n-1}}{(n-1)!} e^{-\beta t} \tag{1}$$

with parameters β and n, where n is the threshold number of hits. The expectation of the distribution is n/β. The distribution reduces to the exponential distribution when $n = 1$; then survival is age independent. With increasing n, survival becomes more age dependent and the variance in survival times decreases. Our model describes adult survival by the Erlang distribution using default parameters derived from our experimental data. We set the life expectancy n/β at 0.5 of the development time (in other words: $\alpha/\lambda = 2$), and the shape parameter n of the distribution, or number of hits, at five. Rosewell and Shorrocks (1987) suggest that survival of adult *Drosophila* in the field is largely age independent. However, their experiments are unlikely to pick up age dependence, because they start with flies of unknown age. Moreover, they captured and recaptured flies on trays of fruit. The flies thus had access to large amounts of food, while our experiments show that survival becomes more age dependent when food is less plentiful (Sevenster and Van Alphen, *in prep.*). In our model, the adult population consists of 100 age classes, which at the end of each time step are multiplied with their age specific survival and moved to the next class. The last age class, which is extremely small, is dropped. The total of all adult age classes is used as the number of adults in equation 2 below.

At every time step a certain quantity of larval food (expressed as K, see below) may or may not become available. The intervals between these reproductive opportunities (hereafter: "intervals") are of random length; each day is a reproductive opportunity with probability 1 over i, which leads to an average interval of approximately i days. This is the *only stochastic* component in the model.

The number of recruits produced at reproductive opportunities by either species is modelled here by Maynard Smith and Slatkin's (1973) equation, which describes many kinds of density dependence (Bellows 1981). The equation is:

$$N_{1r} = \frac{R_1 N_{1a}}{1 + (R_1 - 1)(\frac{N_{1a} + c_{12} N_{2a}}{K_1})^b} \tag{2}$$

where N_{1r} is the number of recruits to the adult population of species 1 at time $t + \alpha_1$, t being the present time and α_1 being species 1's developmental period (see below); N_{1a} is the present number of adults of species 1; N_{2a} is the present number of adults of the competing species 2. The model includes an analogous equation for recruitment of species 2.

R_1 in equation 2 is the number of offspring of species 1 per head and per reproductive opportunity in the absence of any competition. Few data are available on R for species other than *D. melanogaster*. Using Shorrocks and Rosewell's (1986) data on the daily fecundity of temperate species, R_1 and R_2 were set at 15. It should be noted, that daily fecundity of *Drosophila* generally decreases with age (e.g. Ashburner and Thompson 1978), while in our model R is kept constant. However, decreased fecundity at higher ages would have the same consequences as increased mortality at higher ages in our model. Age dependent fecundity could therefore be incorporated into the model by adjusting the adult survival function. Given the short life span and the relatively slow decline in daily fecundity, the necessary adjustment would be minor.

Parameter c_{12} is the competition coefficient. In the present case, we assume that the fast species (with the highest developmental rate) experiences almost no harm from the slow species: $c_{fast,slow}$ is small. The slow species on the other hand suffers seriously from the fast species' presence: $c_{slow,fast}$ is high. Using the range of competition coefficients as found in the literature (Shorrocks and Rosewell 1987), we set $c_{fast,slow}$ and $c_{slow,fast}$ at 0 and 3, respectively. K is the number of adults at which each adult realizes one offspring in the absence of the competing species. This scaling parameter is set at 1000 for both species. The carrying capacity K is kept constant because there is no reason to believe that the size of patches (e.g. fruits or fruiting trees) is correlated with the frequency at which they occur in the *Drosophila* system. Parameter b determines the shape of the competition curve. When b increases from 1, competition changes from contest to scramble. The value of b in *Drosophila* systems varies between 1 and 3 (Shorrocks and Rosewell 1986, 1987) and we set its default in the model at 2.

Juveniles may sit safely in a patch of food or in their pupa, while adults suffer from a bad environment caused by competition or a low abundance of breeding sites. They can buffer the system through a storage effect (Chesson and Huntly 1989). Therefore, the delay of recruitment caused by the developmental period α should be included in the model. Recruits are put in a negative age class and added to the adult population after their developmental period α. The initial values of α for the two species are set at 9 and 15 days to represent the neotropical species *D. willistoni* and *D. sturtevanti* which were used in population cage experiments (Sevenster and Van Alphen, *in prep.*).

If the sum of all individuals in adult and pre-adult age classes drops below one during a simulation, a species is considered to be extinct. Obviously, no run is the same as another due to the stochasticity of the intervals between reproductive opportunities. Therefore, the model is run 100 times for each particular combination of parameters to obtain average figures for every average interval length i from 1 to 20. The parameters of the model are varied one by one in a sensitivity analysis. The parameter values are given in Table 1.

The simulations start with 500 adults of both species in the first age class. They last 100 time steps (days). The relative average numbers of individuals the species are then taken as a measure of their relative performance in an environment characterized by an average interval i. A period of 100 days may not seem very long with respect to the matter of coexistence. However, fruit abundance in the field changes quite rapidly (Foster 1982, Sevenster and Van Alphen *in prep.*) and real populations are not likely to experience the same environmental regime for long. Moreover, it should be noted that in this stochastic model, sooner or later there will always be a run of foodless days long

Table 1. Values of parameters used in the simulation model. Information on the source of the parameter values can be found in the text.

Parameter	Default	Sensitivity analysis
Length of run	100	200
Amount of food K	1000	–
Initial population sizes	500	–
Average interval i	1–20	–
Shape of competitive curve b	2	1, 3
Pre-adult period over life expectancy α/λ	2	1,4
Shape of survival curve n	5	1
Fecundity R fast sp./slow sp.	15/15	5/5, 50/50,25/15
Developmental period α fast sp./slow sp.	9/15	9/10, 14/15
Competition coefficients $c_{fast,slow}/c_{slow,fast}$	0/3	see text
Extra age and species independent daily mortality	0	0.30

enough to drive both populations to extinction. The model should be seen as simulating local populations, in which all individuals experience the presence and absence of food simultaneously. These populations can then be the sources of colonization of other such populations.

Results

Figure 1 presents the outcome of the model as the number of adult individuals of each species after 100 days as a function of the average interval between breeding opportunities i. With both species starting at 500 individuals, the slow species ends up with a larger population than the fast species if the interval i is longer than 6 time steps. In other words, the "inferior competitor" becomes relatively abundant when food is scarce. However, we can not tell whether either species will eventually be excluded without an invasibility analysis (see below).

In the sensitivity analysis, the default parameters of the model are varied to test its robustness (Table 1). In Figures 2, 3, 4, 6 and 7 the outcome of such simulations is presented as the average fraction of adult individuals belonging to the slow species after 100 days. This method does not show the actual population sizes after 100 days, but indicates the performance of one species relative to the other. In all figures, the solid line represents the default parameter combination, while the broken lines indicate the variations. Some of the curves are far from being smooth for long intervals. Due to the exponential distribution of interval lengths i, the variance of i is proportional to the square of i. Runs with the same parameter values will therefore show more variation at higher i's. Moreover, the average population size at i's near 20 will be determined by a few runs without extinctions.

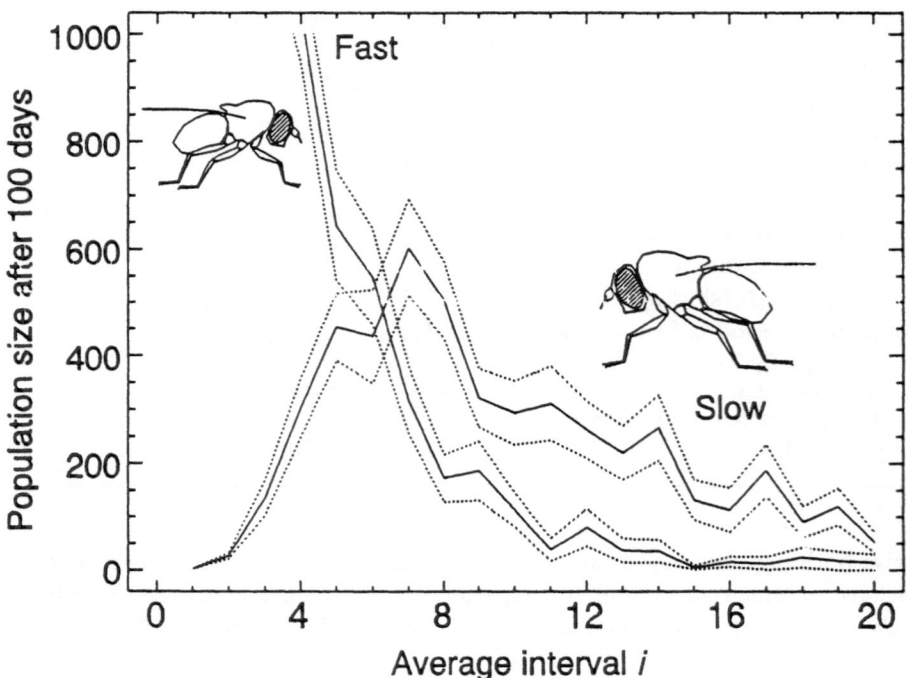

Figure 1. Population size of fast and slow species against average interval length (means and standard errors of 100 simulations per value of i). The population size of the fast species is 2212 for i equals 1. For values of i larger than 7, the slow species (poor larval competitor) performs better than the fast species.

The trends seen at 100 days are accentuated when the length of the simulation is increased to 200 time steps. With both species starting at 500 adults, the fast species becomes fixed for i equals 1 or 2 and the slow species becomes fixed when i is 12 or up. The type of competition b does not change the default curve, although scramble competition (b equals 3) seems to create more variation than contest competition (b equals 1).

The output of the model changes little or not at all, when the number of offspring per reproductive opportunity per head, R, is varied in the same way for both species. However, daily fecundity may not be independent from the trade off between developmental period and adult survival. In fact, fast species may have higher daily fecundities than slow species (Hoffmann and Parsons 1989). Therefore we present a series of simulations with R of the fast species increased by about 70% to 25. The R of the slow species was kept at 15. Figure 2 shows that the outcome of competition is barely influenced by this change. Increased competition compensates the increased fecundity of the fast species, especially at the high densities found at low i's (cf. Figure 1).

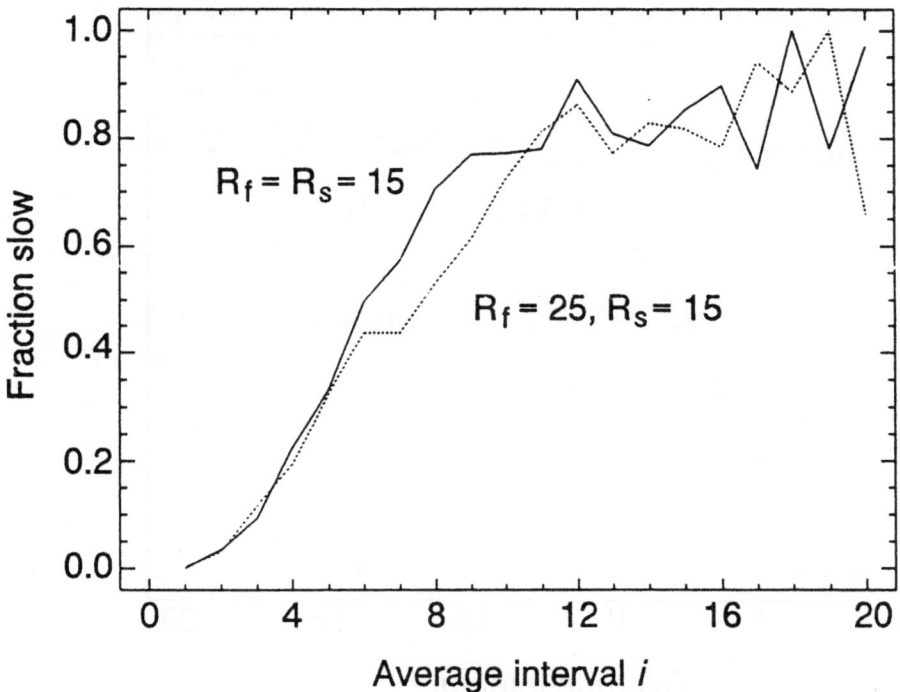

Figure 2. The outcome of competition when the reproductive output per time step, R, increases with developmental rate. If R of the fast species, R_f, is increased from 15 to 25, the outcome of competition barely changes.

Life expectancy λ can be measured in the lab, but little accurate information exists about natural situations (Shorrocks and Rosewell 1987 summarize the survival data available on temperate species). Therefore, the robustness of the model with respect to the α/λ constant must be considered carefully. Figure 3 shows that the relative performance of the slow species improves considerably with increasing α/λ, i.e. with a decreasing life expectancy. Natural life expectancies will be such that α/λ is always higher than 1 (Shorrocks and Rosewell 1987, Charnov and Berrigan 1990).

In Figure 4 the shape parameter n of the adult survival function is varied. The default value is $n = 5$, modelling age dependent survival. When $n = 1$, survival is age independent, i.e. exponential. This takes away much of the advantage of being long lived, because there is no period with low mortality to be extended beyond i by shifting the balance in the trade off. Still the slow species performs best for i's above 12 days.

Experimental results have demonstrated the trade off between developmental rate and adult mortality under laboratory conditions (Sevenster and Van Alphen, *in prep.*). In the field, however, predation and other sources of "external" mortality could be similar for different species and high relative to the "intrinsic" mortality discussed so far. On the other hand, flies selected for desiccation resistance (Hoffmann and Parsons

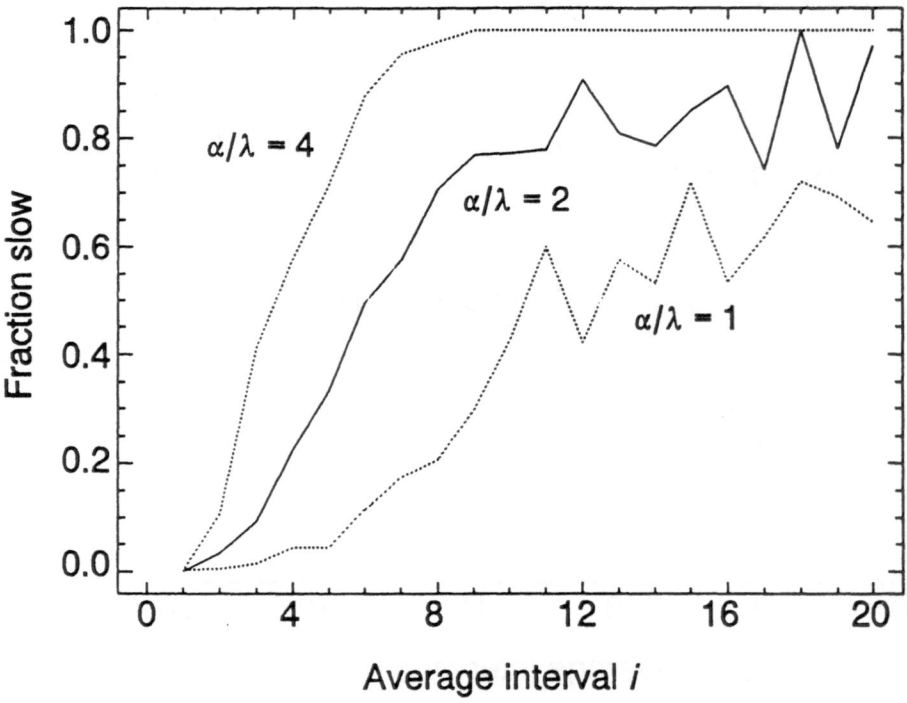

Figure 3. The influence of the ratio of developmental period α to life span λ. This ratio is the same for both species. If α/λ increases, the relative performance of the slowly developing species improves.

1989) or postponed senescence (Service 1987) are less active than controls, which could make them less accident prone. Nevertheless, we present a series of simulations in which the adult mortality figures derived from the experiments are augmented with a high extra mortality of 0.30 per day, independent of age and species. The latter figure is nearly as high as the average total mortality in the field data compiled by Shorrocks and Rosewell (1987). This parameter combination may be seen as a particularly rigorous test of our hypothesis. The survival curves of the species are lowered dramatically (Figure 5), but the consequences for the relative performance of the competing species are only slight (Figure 6), due to the fact that the relative difference in survival at ages above 4 days remains large. The drop of the curve for i equals 19 and 20 is generated by stochastic variation between simulations as explained in the second paragraph of the results section.

    The present model is based on the notion that the developmental period α determines the competitive ability of a larva. Therefore, in order to investigate the consequences of the magnitude of the difference in strategy, we have to rewrite the competition

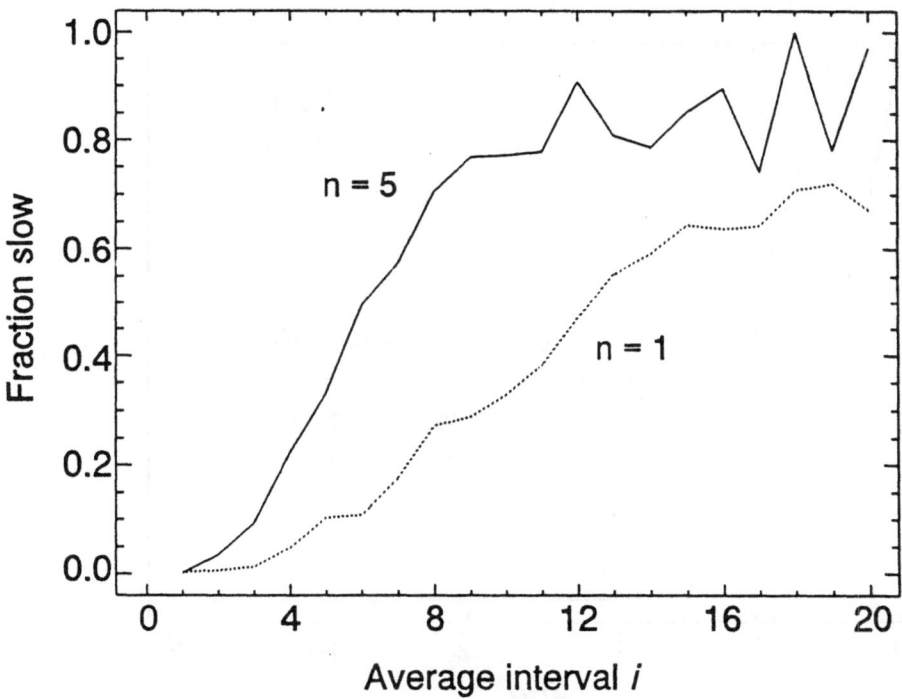

Figure 4. The age dependence of adult mortality ($n = 5$) greatly contributes to the success of the slow species. If adult mortality were age independent ($n = 1$), the performance of the slow species would be much reduced.

coefficients c as a function of that difference:

$$c_{pq} = \left(\frac{\alpha_p}{\alpha_q}\right)^a \tag{3}$$

in which a lower value of a means a smaller difference in larval competitive ability. Consequently, low a's lead to a better performance of the slow species. Parameter a is set at 2 ($c_{\text{fast,slow}} = 0.36$ and $c_{\text{slow,fast}} = 2.78$ for the default developmental periods α of 9 and 15 days) to analyze what happens when the difference in α gets smaller. The results remain virtually unchanged, unless the difference in strategy is much reduced. In that case, the fraction of slow adults after 100 time steps approaches 0.5 (Figure 7). The curves for $\alpha_f = 9$, $\alpha_s = 10$ and for $\alpha_f = 14$, $\alpha_s = 15$ show that the percentage difference rather than the absolute difference in α determines the advantage of a species.

So far, the model showed that the fast species performs best when breeding opportunities are frequent, and that the slow species performs best when breeding opportunities are rare. This result is generally robust against changes in parameter values, although the switching point may shift. A most interesting question now is whether the differences between fast and slow species could lead to stable coexistence. Species will coexist

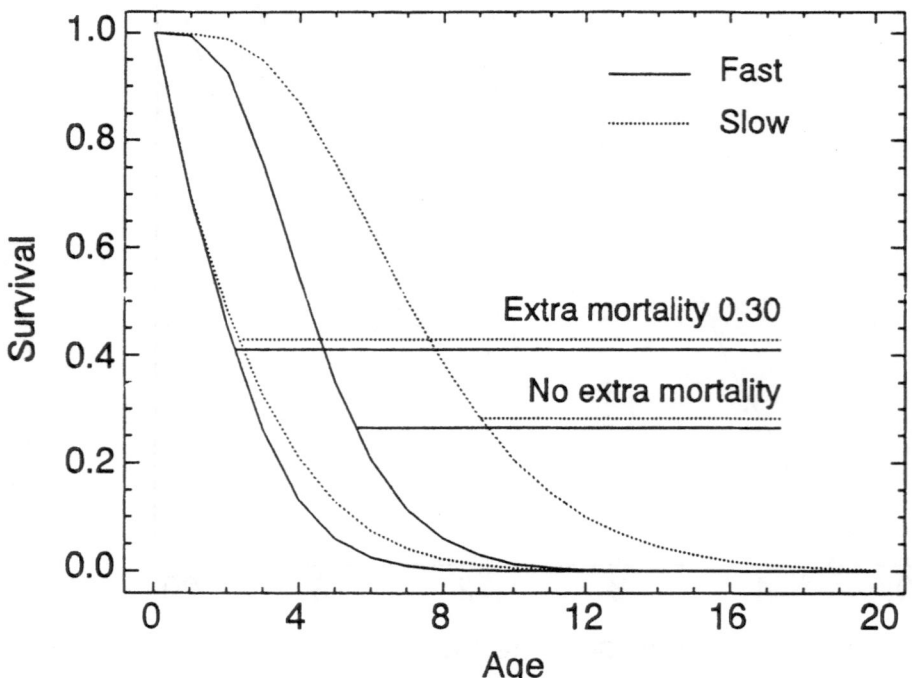

Figure 5. Survival curves with and without an extra age and species independent adult mortality of 0.30 per time step. Although this extra mortality drastically reduces survival, a considerable difference between fast and slow species remains for ages above 4 time steps (days).

stably, if both are able to invade in a system where the other species is established. We apply this mutual invasibility criterion to the model in the following way. We start the system with one adult of one species and let the model run for 100 time steps. The average population size has then reached a plateau, and this species is assumed to be "established," although the stochasticity of the environment will prevent a constant population size. Now one adult of the other species invades and the model runs another 100 time steps. The simulations were repeated 200 times for every value of the average interval length i. In these simulations the extinction criterion is dropped, since the interest does not lie in the fate of one small local population. We are interested here in the expected long term success of an invader in a temporally variable, but spatially homogeneous system. In order to analyze this situation correctly, we have to take the geometric mean of the result of single invasions (see discussion).

The invading species starts with a population size of one. If the geometric mean population size 100 time steps after invasion is larger than one, an invasion is expected to be successful. The mutual invasibility analysis shows that the two species are able to invade each other for average interval lengths i around 5 days (Figure 8). The present

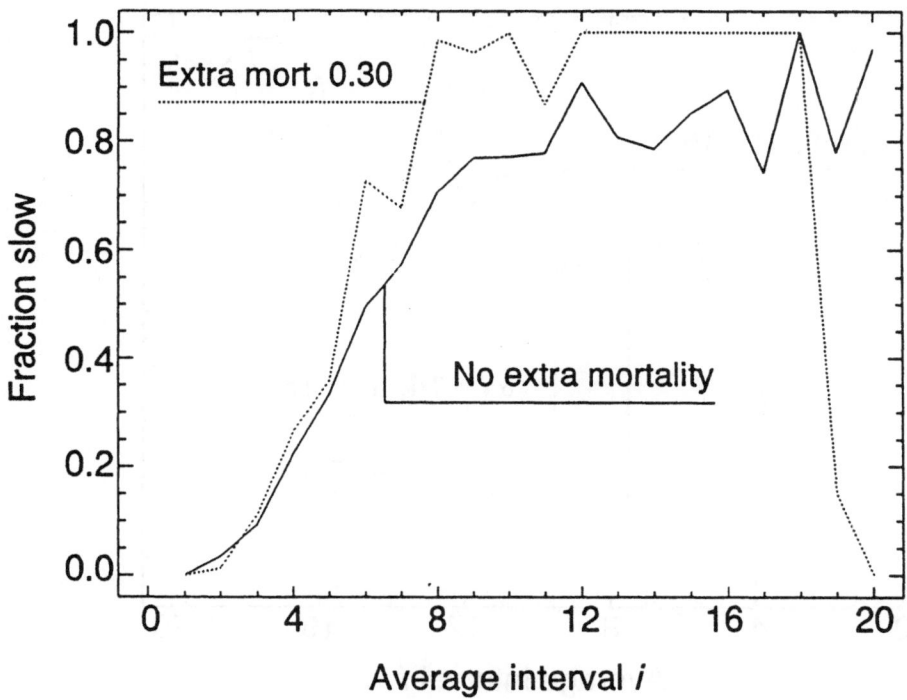

Figure 6. The outcome of competition when an extra age and species independent adult mortality of 0.30 per time step is applied. The relative performance of the slow species seems to improve somewhat with this extra mortality. The drop in the line for extra mortality at *i* equals 19 and 20 is caused by stochastic variation between runs (see text).

analysis demonstrates that temporal stochasticity of the environment may result in stable coexistence of the two species in the model, even if the environmental regime (i.e. average interval length *i*) does not change.

Discussion

Competition experiments suggest that among *Drosophila* species superior and inferior species can be distinguished (e.g. Gilpin et al. 1986). It seems odd that competitively superior and inferior species would persist side by side. The present results demonstrate that the benefits of being an inferior larval competitor emerge under conditions of low food abundance, while in the usual competition experiment food is continuously present. Our conclusion is that long lived species, that are bad larval competitors, are superior when the probability of encountering reproductive opportunities is low. The sensitivity analysis demonstrates that the model is qualitatively robust against changes in parameter values. Additional analyses demonstrate that the distribution of interval

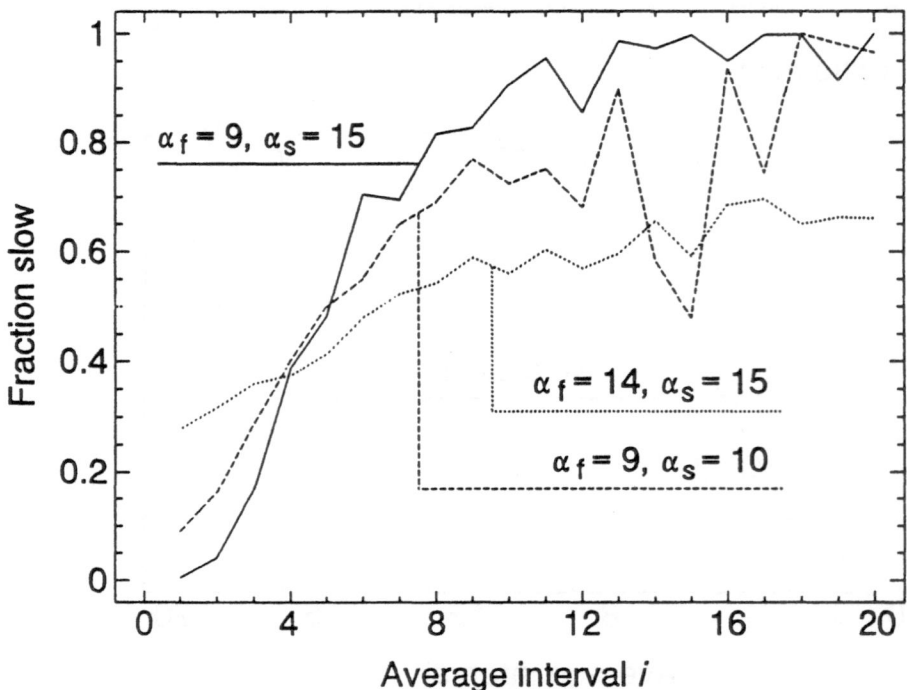

Figure 7. When the magnitude of the difference in strategy is decreased to one time step (day), the difference in performance of the two species is reduced. The strategy is characterized by the developmental period α (α_f and α_s being the developmental period of the fast and the slow species, respectively). See text for further discussion.

lengths does not change the qualitative conclusions of the model. However, the model generates synchronization effects if the variance of interval lengths is much reduced. In that case, a species benefits greatly from intervals that equal or slightly exceed its generation time.

Our mutual invasibility analysis demonstrates that the trade off between larval competitive ability and adult survival enhances coexistence when breeding opportunities occur intermittently. According to Chesson and Huntly (1988, 1989) and Chesson and Rosenzweig (1991), switching dominance and a subadditive interaction between the environment and competition are necessary to stabilize coexistence in variable environments. These conditions are met in the *Drosophila* system, where the impact of larval competition decreases with decreasing environmental favorability. In our model, the carrying capacity per reproductive event is constant. Nevertheless, the competing populations become smaller when the environment becomes less favorable (i.e. when the feeding interval becomes longer; see Figure 1), and thus the intensity of larval competition must be reduced. This result is confirmed by our field data (unpubl.) on the frugivorous *Drosophila* from Panama: in general, fewer flies emerge per fruit when fruits

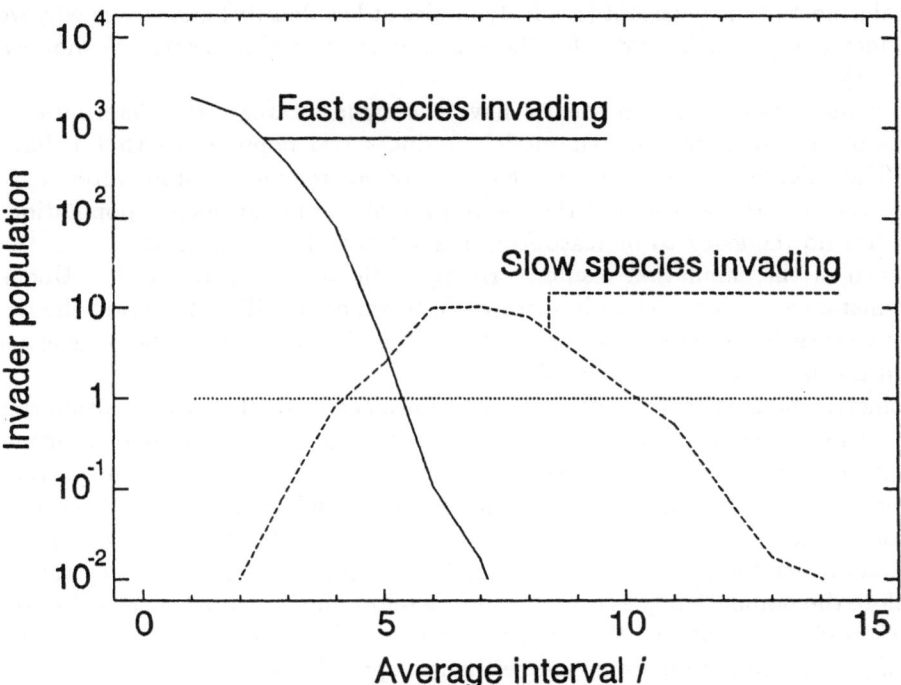

Figure 8. The geometric mean population size of the invading species 100 days after invasion, in a temporally stochastic system in which the other species is established. Values greater than unity indicate that invasions can be expected to be successful.

are scarce than when fruits are abundant. In most organisms competition would intensify when food is scarce. In *Drosophila* the situation is different, because the generation time is short relative to the time scale of fluctuations in resource abundance.

Our results show that two species may coexist stably in an environment where breeding opportunities occur stochastically in time. The question arises, however, whether this situation is evolutionarily stable. It is well possible that any given average interval between breeding opportunities has its evolutionarily stable strategy, which can not be invaded. A certain difference in strategy seems to be necessary to let one species benefit from the weaknesses of the other (Figure 7).

We used the geometric mean result of simulated invasions to assess invasibility. We will explain why this is correct, even though there is no temporal connection between invasion events (Yoshimura and Clark 1991). The simulations often show that the average population size of the invader increases at first (roughly during the first 100 time steps) and decreases later. The reason is that the sampling path of the population size includes many high peaks shortly after the start of the simulation, while such peaks become increasingly rare later. We choose to investigate the expected *long term* success of an invader. That perspective seems to be most relevant to issues such as stable polymorphisms or stable coexistence of species. Therefore, we should not look at the average population size shortly after the invasion. Instead, we must examine whether the invading species is able to recover *repeatedly* from low densities. Here is the sequential aspect of the invasions (read: dynamics at low density) and this is why we used the geometric mean as indicator for the expected success of invasions (Yoshimura and Clark 1991).

In the extreme case that two species have the same strategy, the deterministic nature of population dynamics in our model produces two populations that behave identically. The *relative* numbers of two identical species remain constant after invasion, because the species behave as if they were part of one homogeneous population. The invader has no tendency to increase from low relative densities. In this case, the geometric mean of the simulation results correctly indicates *exact neutrality*. Under natural circumstances such a system is susceptible to random drift. Therefore, the coexistence of two identical species in a temporally stochastic, but spatially homogeneous system would not be likely to last indefinitely.

The geometric mean is heavily influenced by values near zero. If any simulation run ends in the extinction or near extinction of the invading species, the geometric mean indicates that the invader would not be successful in the long run. This may seem incorrect, because all other runs could produce strong population growth. However, any extinctions indicate a finite chance that the invading population does not survive low densities and, therefore, persistence would not be lasting indefinitely. It may take a long time before the population goes extinct, and therefore the geometric mean criterion may be considered conservative. However, we believe it is the best criterion for the persistence of populations in environments without a spatial structure.

In our model, sooner or later there will always be a run of foodless days long enough to drive both species to extinction. This seems to be at odds with the long term approach of the invasibility analysis. However, the invasibility analysis includes the environment. This explains why the slow species can not persist indefinitely for average intervals longer than 10 days. The slow species is not outcompeted by the fast species in that case, but the environment is too poor to allow indefinite persistence. We do not think that it would be biologically relevant to try and circumvent this "artifact" of the model.

The arithmetic mean is appropriate when the system consists of many "parallel" sub-systems that are independent for the length of the simulation run. To explain this, we will interpret our results as if they were meant to simulate a patchy system. Assume that the system consists of 200 patches containing individuals of one species. These patches independently experience 100 time steps of stochasticity prior to the invasion of the second species. As a result, the local population size of the resident at the time of invasion varies between patches. Now, propagules (adult flies) of the invading species enter the patches at very low densities ($N = 1$ in the model). The patches experience independent stochasticity for another 100 time steps. At the end of these 100 time steps, the local populations merge into one metapopulation, and we want to know the number of propagules of the invading species that can start new local populations. If the arithmetic mean of the number of propagules per patch exceeds the density at invasion, the total number of propagules in the metapopulation has increased and we

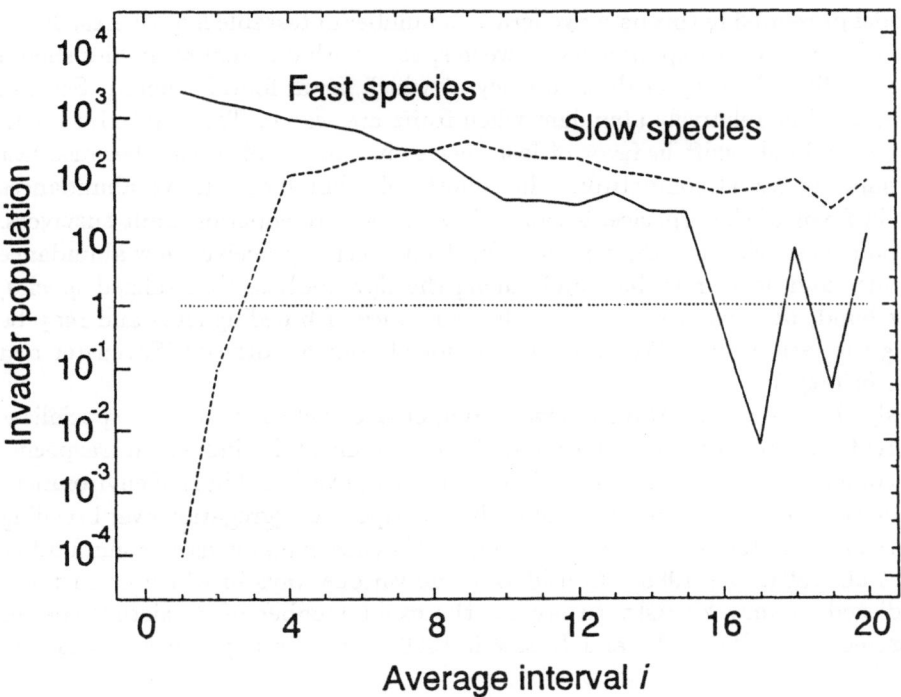

Figure 9. The arithmetic mean population size of the invading species 100 days after invasion.

expect the invading species to persist in this patchy system. Figure 9 shows that in such a patchy system, a fast and a slow species coexist stably over a much larger range of average intervals i, than in the spatially homogeneous system (Figure 8). Apparently, the success of invaders in patches with few competitors more than compensates for the losses in other patches (Comins and Noble 1985).

This scenario for a patchy system is not at all unrealistic for *Drosophila*. In frugivorous species, concentrated populations exist under fruiting trees. When the trees cease to produce fruits, the local populations dissolve and then concentrate under new fruiting trees. However, our simulations show that not only species with different life histories, but even identical species can invade each other in such patchy systems. The combination of spatial and temporal variability thus leads to stable coexistence of competitors without any differentiation. This conclusion reminds one of the Aggregation Model of coexistence in which species compete on ephemeral patches and then redistribute themselves in an aggregated fashion over new patches in the next generation. Under those circumstances, species may coexist without any differentiation (Atkinson and Shorrocks 1981; Hanski 1981; De Jong 1982; Ives and May 1985). The conclusion also agrees with Chesson's (1985a) finding that pure temporal variation is a less effective promotor of coexistence than spatial variation for short lived animals. Undoubtedly, the interpretation of our simulations for patchy systems is relevant. However, we used the geometric mean criterion because we are interested in what a difference in life history may do for coexistence. Differences in life history are apparently less crucial in spatially structured systems.

The model presented in this paper generates a number of testable hypotheses. First, the outcome of competition experiments between species with different strategies should depend, in a predictable way, on the frequency at which larval food is offered. Second, slow species should be relatively abundant when fruits are scarce. Third, relative abundances should gradually shift in favor of fast species in cohorts of larvae, because fast species are superior larval competitors. In cohorts of adult flies relative abundances should shift in favor of slow species, because slow species are superior adult survivors. Fourth, specialized species, breeding in but a few fruit species, perceive a low abundance of breeding sites and hence may be found among the slow species. Generalized species, on the other hand, perceive a relatively high abundance of breeding sites and may be found among the fast species. We have support for all four hypotheses (Sevenster and Van Alphen, *in prep.*).

Obviously there are other ways to evade competitive exclusion, such as specialization on different resources (e.g. MacArthur 1972), environmentally induced intraspecific variation in competitive ability (Begon and Wall 1987) and various kinds of environmental heterogeneity (Chesson 1985a). Undoubtedly, intraspecific aggregation over breeding sites is a particularly potent mechanism allowing coexistence in many insect communities (Shorrocks et al. 1984, Ives 1988). Considering the various ways in which coexistence may be mediated, it may be risky to predict the exact number of coexisting species based on aggregation (Shorrocks and Rosewell 1986, 1987) or any other mechanism alone.

Acknowledgement

We wish to thank Kees Bakker, Steven R. Beissinger, Bert van den Bergh, Thomas J. DeWitt, Gerard Driessen, Ian Hardy, Allen Herre, Arne Janssen, Lex Kraaijeveld, Hans Metz, Janine Pijls and Jin Yoshimura for their comments on this paper. The first author was partly funded by grant W84-262 of the Netherlands Foundation for the Advancement of Tropical Research and a short term fellowship from the Smithsonian Tropical Research Institute, Panama.

References

Ashburner, M. and J.N. Thompson Jr. 1978. The laboratory culture of *Drosophila*. In M. Ashburner and T.R.F. Wright (Eds.) *The Genetics and Biology of Drosophila 2a*. Academic Press, London. pp. 1–109.

Atkinson, W.D. 1979. A comparison of the reproductive strategies of domestic species of *Drosophila*. *J. Anim. Ecol.* 48: 53–64.

Atkinson, W.D. and B. Shorrocks. 1981. Competition on a divided and ephemeral resource: a simulation model. *J. Anim. Ecol.* 50: 461–471.

Bakker, K. 1961. An analysis of factors which determine success in competition for food among larvae of *Drosophila melanogaster*. *Arch. Neerl. Zool.* 14: 200–281.

Bakker, K. 1969. Selection for rate of growth and its influence on competitive ability of larvae of *Drosophila melanogaster*. *Neth. J. Zool.* 19: 541–595.

Begon, M. and R. Wall. 1987. Individual variation and competitor coexistence: a model. *Funct. Ecol.* 1: 237–241.

Bellows, T.S. 1981. The descriptive properties of some models for density dependence. *J. Anim. Ecol.* 50: 139–156.

Charnov, E.L. and D. Berrigan. 1990. Dimensionless numbers and life history evolution: age of maturity versus adult life span. *Evol. Ecol.* 4: 273–275.

Chesson, P.L. 1985a. Coexistence of competitors in spatially and temporally varying environments: a look at the combined effects of different sorts of variability. *Theor. Popul. Biol.* 28: 263–287.

Chesson, P.L. 1985b. Environmental variation and the coexistence of species. In J. Diamond and T.J. Case (Eds.) *Community Ecology*. Harper and Row, New York. pp. 240–256.

Chesson, P. and N. Huntly. 1988. Community consequences of life history traits in a variable environment. *Ann. Zool. Fennici* 25: 5-16.

Chesson, P. and N. Huntly. 1989. Short term instabilities and long term community dynamics. *Trends Ecol. and Evol.* 4: 293–298.

Chesson, P. and M. Rosenzweig. 1991. Behavior, heterogeneity, and the dynamics of interacting species. *Ecology* 72: 1187–1195.

Comins, N.R. and I.R. Noble. 1985. Dispersal, variability, and transient niches: species coexistence in a uniformly variable environment. *Amer. Nat.* 126: 706–723.

De Jong, G. 1976. A model of competition for food. 1. Frequency-dependent viabilities. *Amer. Nat.* 110: 1013–1027.

Foster, R.B. 1982. The seasonal rhythm of fruitfall on Barro Colorado Island. In E.G. Leigh, A.S. Rand and D.M. Windsor (Eds.) *The Ecology of a Tropical Forest*. Smithsonian Institution Press, Washington. pp. 151–172.

Gilpin, M.E., M.P. Carpenter and M.J. Pomerantz. 1986. The assembly of a laboratory community: multispecies competition in *Drosophila*. In J. Diamond and T.J. Case (Eds.) *Community Ecology*. Harper and Row, New York. pp. 23–40.

Hanski, I. 1981. Coexistence of competitors in patchy environment with and without predation. *Oikos* 37: 306–312.

Hardin, G. 1960. The competitive exclusion principle. *Science* 131: 1292–1297.

Harvey, P.H., M.D. Pagel and J.A. Rees. 1991. Mammalian metabolism and life histories. *Amer. Nat.* 137: 556–566.

Hoffmann, A.A. and P.A. Parsons. 1989. An integrated approach to environmental stress tolerance and life history variation: desiccation tolerance in *Drosophila*. *Biol. J. Linn. Soc.* 37: 117-136.

Ives, A.R. 1988. Covariance, coexistence and the population dynamics of two competitors using a patchy resource. *J. Theor. Biol.* 133: 345–361.

Ives, A.R. and R.M. May. 1985. Competition within and between species in a patchy environment: relations between microscopic and macroscopic models. *J. Theor. Biol.* 115: 65–92.

MacArthur, R.H. 1972. *Geographical Ecology*. Harper and Row, New York.

Maynard Smith, J. and M. Slatkin. 1973. The stability of predator-prey systems. *Ecology* 54: 384–391.

Nunney, L. 1983. Sex differences in larval competition in *Drosophila melanogaster*: the testing of a competition model and its relevance to frequency-dependent selection. *Amer. Nat.* 121: 67–93.

Richter, O. and D. Söndgerath. 1990. *Parameter Estimation in Ecology*. VCH, Weinheim.

Rosewell, J. and B. Shorrocks. 1987. The implications of survival rates in natural populations of *Drosophila*: capture-recapture experiments on domestic species. *Biol. J. Linn. Soc.* 32: 373–384.

Service, P.M. 1987. Physiological mechanisms of increased stress resistance in *Drosophila melanogaster* selected for postponed senescence. *Physiol. Zool.* 60: 321–326.

Sevenster, J.G. and J.J.M. van Alphen. *in prep*. A life history trade off in *Drosophila* species and community structure in variable environments.

Shorrocks, B., J. Rosewell, K. Edwards and W. Atkinson. 1984. Interspecific competition is not a major organizing force in many insect communities. *Nature* 310: 310–312.

Shorrocks, B. and J. Rosewell. 1986. Guild size in Drosophilids: a simulation model. *J. Anim. Ecol.* 55: 527–541.

Shorrocks, B. and J. Rosewell. 1987. Spatial patchiness and community structure: coexistence and guild size of Drosophilids on ephemeral resources. In J.H.R. Gee and P.S. Giller (Eds.) *Organization of Communities Past and Present*. Blackwell Scientific Publications, Oxford. pp. 29–51.

Strong, D.R., D. Simberloff, L.G. Abele and A.B. Thistle (Eds). 1984. *Ecological Communities: Conceptual issues and the Evidence*. Princeton University, Princeton.

Treveleyan, R., P.H. Harvey and M.D. Pagel. 1990. Metabolic rates and life histories in birds. *Funct. Ecol.* 4: 135–141.

Yoshimura, J. and C.W. Clark. 1991. Individual adaptation in stochastic environments. *Evol. Ecol.* 5: 173–192.

THE EQUILIBRIUM DISTRIBUTION OF OPTIMAL SEARCH

AND SAMPLING EFFORT OF FORAGING ANIMALS

IN PATCHY ENVIRONMENTS

Dan Cohen

Department of Evolution, Systematics and Ecology
The Hebrew University
Jerusalem 91904, Israel

Abstract. The optimal allocation of time and effort by foraging animals for searching, sampling and learning the distribution of food in patches is modelled. The optimal effort maximises the net gain, which is the difference between the benefit and the cost of searching. The optimal search effort increases as a function of the variance of quality between the patches, and the turnover of new patches. It decreases as a function of the cost of searching, e.g. the movement cost between the patches.

'The variance of quality in the patches decreases as a function of the total search effort by all the foragers. A joint stable equilibrium of search effort and patch variance is reached when the variance reaches the level which is generated by the search effort which is optimal for all the foragers. The joint equilibrium search effort is a decreasing function of the population density, and of the search cost. It is an increasing function of the inherent variance in patch quality, and of the patch renewal rate. The equilibrium variance in patch quality is a decreasing function of the populaton density of the foragers, and of the search cost.

A rare type with a different optimal search as a function of the variance, behaves according to the variance equilibrium with the common type. The optimal search effort of rare types diverges from that of the common type more than it would if they were on their own. In heterogeneous populations with several different types, the equilibrium search effort of each type is the optimal search effort at the variance generated by the total search effort of all the types.

Coexistence is possible between species with high information gathering and high energy reqirements, which utilise the richer newly discovered food sources, and species with low information gathering and low energy demands, which utilise poorer depleted food sources.

A small average number of foraging visits per patch generates a stochastic distribution of patch quality even in inherently uniform patches. In such cases, the equilibrium search effort and patch variance may be high, and may be determined entirely by this stochastic generation of variance.

I. Introduction

Foraging strategies specify the allocation of foraging time and effort to environmental patches of different types, the choice of alternative search and sampling strategies, and the choice of dietary items as they are encountered. For general reviews see Krebs and Davies (1984), Pyke (1984), Kamil, Krebs and Pulliam (1987).

Optimal foraging theories construct models and predictions for the short term dynamics and equilibrium of foraging behaviour, and for the long term evolutionary changes, over a range of environmental conditions and behavioural and physiological constraints.

In general, a foraging animal has to decide between the following alternatives:
    a. to continue feeding in a known patch;
    b. to leave a patch and return to a previously visited patch;
    c. to search for new patches.
If an animal has complete information about the distribution of resources in the patches, a reasonable optimal decision rule for leaving a patch is the Marginal Value Theorem (Charnov 1976). According to this rule, an animal should leave a patch when the expected rate of the net local resource acquisition in the patch falls below the average expected rate in the habitat.

Since the foraging activity depletes the local level of resources in a patch, and thus reduces the expected fitness of foragers in that patch, a stable equilibrium (ESS) distribution of foraging activity in a population of foragers is expected to equalise the fitness of all the animals with the same fitness functions and physiological constraints. With complete information, and with equal movement cost between the patches, the ESS distribution of foraging activity is also expected to lead to the Ideal Free Distribution of equal average quality in all the exploited patches (Fretwell 1972).

The role of information in optimal foraging strategies in patchy and varying environments. A very important component of the foraging decisions in natural environments is the information that the animal has about the conditions in the currently exploited patch, and about the distribution of food quality and quantity in other available patches (e.g. Pyke 1981, Green 1984, 1987, Tamm 1987, Waddington 1983, Real 1983, Kacelnik and Krebs 1985, Clark and Mangel 1984, Mangel 1990). Animals compete for access to information and for the exploitation of newly discovered resources (Caraco 1987).

Therefore, the decision rules which form realistic foraging strategies must specify the allocation of time and effort to searching and sampling, the rule for leaving a patch, and the rule for choosing where to go next, as functions of the presently available information (e.g. Cook and Hubbard 1977, Waage 1979, Lima 1984, McNamara and Houston 1985, Mangel 1990, Krebs and Inman 1992).

Optimal search models have been developed for analogous economic activities, e.g. natural resource utilisation (Mangel 1985, Mangel and Clark 1986), searching for jobs (McKenna 1985), and shopping for the best price (Burdett and Judd 1983). The importance of the limiting physiological and perceptual mechanisms of learning and neural constraints has been pointed out by Houston, Kacelnik and McNamara (1982), Menzel (1985), Gould (1984) and Real (1992).

Clearly for animals with complete information, there is no advantage for any effort invested in collecting and processing information.

Optimal foraging in naturally varying or heterogeneous environments can be divided therefore in principle into two stages:

1. Gathering information about the probability distribution of environmental conditions by searching, sampling and learning.
2. Making optimal foraging decisions based on the available information.

Thus the optimal foraging strategies have to take into account also the costs of searching, sampling and learning, and the benefits of the additional information contributed by the sampling and searching effort. The optimal search effort is expected to maximise the average net gain in fitness as a function of the search effort.

Note however that the same behavior often contributes to the simultaneous collection of both food and information. In these cases, there is no separate allocation of effort for searching, which cannot therefore be distinguished or measured.

In other cases, which I shall discuss in this paper, it is possible to distinguish between collecting food and gathering information about food. In these cases, it is possible and necessary to model and analyse the contribution of search effort to fitness.

In this paper I present a simple model for the optimal allocation of effort for searching and sampling new patches, for gathering information about the distribution of food quality and quantity in the patches and about the associated costs of movement between the patches. The model is analogous to models of the role of information in economic systems (e.g. McCall 1970, Burdett and Judd 1983).

The model is developed at three levels of increasing complexity and interactions:

A. The first model assumes that the distribution of patch quality is constant and allows the derivation of the optimal search strategy for individual animals.
B. The second set of models considers the effect of the total search effort in the population on the variation in patch quality. Using these models, a joint equilibrium distribution of patch quality and searching and sampling effort can be derived.
C. In the third set of models, I consider heterogeneous populations of the same or of different species. A joint equilibrium of search effort and of the variation of patch quality is then derived for such populations.

II. A model for the optimal search effort for individual animals

Assumptions about the distributions of food and information in the environment.

  1. Food is distributed in patches with a significant cost of movement between them.
  2. Food supply in the patches changes by time-varying or stochastic processes. New patches may appear and old patches may disappear with certain probabilities.

In general, there is always some uncertainty for each animal about both the local conditions and the overall distribution of food in the patches, which is contributed by two components:

  1. The appearance and disappearance of patches and/or the improvement or depletion of some of the existing patches.
  2. An additional component of a subjective uncertainty and lack of information of newly born or newly arrived animals which have not yet searched and sampled a representative fraction of the patches.

I assume that a searching animal has to interrupt feeding in its known patch or patches and to spend time and effort searching and sampling other patches.

Two types of models may be considered:

  1. In a simple model we may assume that a searching animal chooses a predetermined search time, and starts feeding in the best patch that it finds during this search period.
  2. In a more flexible model we may assume that a searching animal chooses a combination of time and patch quality thresholds above which it stops searching and starts feeding in a new patch.

For the sake of simplicity I assume in my model that the search strategy has predetermined search time and effort, S.

A Cost-benefit model for optimal search. Search involves both benefits and costs which are measured in units of fitness. I assume that the optimal search effort for any individual maximises the average or expected net gain in fitness under any particular conditions.

The cost function per unit of search effort during foraging, $C(S)$, is independent of the patch quality distribution. The cost is an increasing function of the energy costs of movement between patches. The cost for the searching animal includes also the cost of not feeding in its known patch during the search, and the risks of predation and other hazards. The cost function $C(S)$ has an increasing slope, because food deprivation usually has an increasing cumulative cost for each additional unit of nonfeeding search time.

A well fed animal has a lower cost function per unit food deprivation than a starved animal because it suffers a lower reduction in fitness by the same amount of food deprivation.

On the other hand, an animal in a good patch loses more potential food intake per unit nonfeeding search time than an animal in a poor patch. However, because animals in good patches are likely to be better fed than animals in poor patches, these two opposing factors complicate the effect of patch quality on the cost function and on the optimal search effort.

Similarly, a young inexperienced animal, or a newcomer to the area, is expected to have a lower cost function, because it has not yet discovered the good food patches in the area and thus feeds in patches with a lower average quality than do experienced animals. It may therefore lose less food intake by allocating time for searching.

The benefit function, $B(S)$, is defined as the increase in expected highest feeding fitness that a searching animal can find as a function of the search effort. It can be defined by the average quality of the best patch encountered by the animal during any search period.

A. In general, $B(S)$ is an increasing function of the variance in the quality of the patches over the range of acceptable quality, because a larger variance provides a larger potential reward for the same search effort. Clearly, for example, searching provides no gain if all the patches have the same quality.

B. Assuming that an animal can remember and return to its currently exploited food patch or resource, the probability of finding a better patch and the expected benefit decrease as a function of the quality of the current patch.

C. The benefit function $B(S)$ has a decreasing marginal return per unit search effort. This is because the probability of finding increasingly better patches decreases with search time for any distribution of patch quality.

In the Appendix I give a mathematical derivation of the benefit function of search effort and its probability distribution.

The optimal level of search effort for any individual animal maximises the average net gain in fitness, i.e. the difference between the benefit and the cost functions $W(S) = B(S) - C(S)$. Since $C(S)$ is concave, and $B(S)$ is convex over the relevant range, there is only one optimal intermediate S, $0 < S^* < T$, where T is the total available time, where the first derivatives at $B(S^*)$ and $C(S^*)$ are equal. S^* may be zero or T if either extreme point maximises the net gain $W(S) = B(S) - C(S)$.

As mentioned above, the benefit of acquiring information for any one animal is contributed by two components:

1. The appearance or disappearance of patches and/or the improvement or depletion of the existing patches, which generates new uncertainty and provides the net gain for searching by established and experienced animals.

2. An additional component of subjective uncertainty and lack of information of newly born and newly arrived animals which have not yet discovered the good patches. The fitness of these inexperienced animals is lower because of their lower energy gain, but they are expected to have a higher optimal search effort because of their lower $C(S)$ cost function and their expected larger net gain by searching. This expected heterogeneity of optimal search effort in a population has important implications for the ESS levels of search effort in heterogeneous populations (see below).

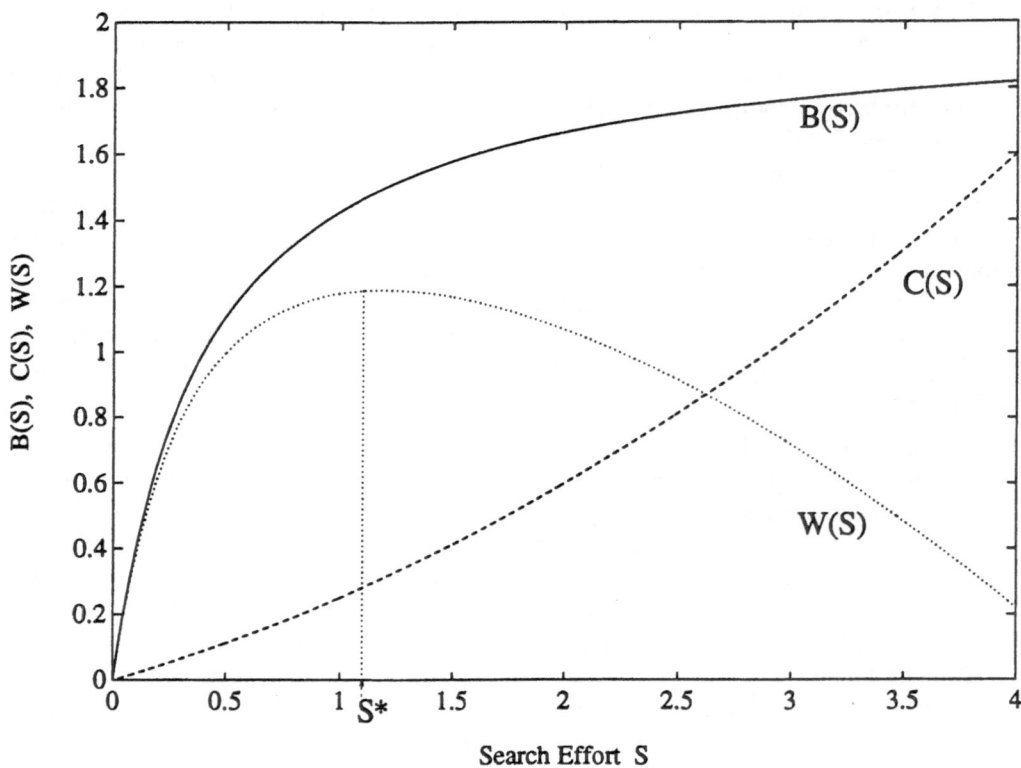

Figure 1. A schematic illustration of the benefit and cost functions of the search effort, $B(S)$ and $C(S)$, and the optimal search effort, S^*, which maximises fitness $W(S) = B(S) - C(S)$, in a uniform population. (Arbitrary fitness and effort units.)

See Figure 1 for an illustration of the optimal search effort in a uniform population with given benefit and cost functions.

III. Modelling the equilibrium search effort in a population: The effect of the search effort on the variance of quality of patches or resources and the resulting equilibrium

Searching animals are more likely to stay and feed longer in the better quality patches, and thereby increase the local feeding intensity in such patches. Similarly, poorer patches are more likely to be abandoned or avoided by searching animals, so that feeding intensity in these patches will decrease.

Possingham (1988) has also shown that more systematic foraging reduces the variance of resources in continuously renewed patches. He applied these results to nectar distribution in flowers, but without explicitly considering the optimal behaviour or information status of the pollinators.

Clearly, if more animals are searching, or if the same number of animals allocate more time for searching, more animals will move into better patches and more animals will leave the lower quality patches.

Since food availability is usually a decreasing function of the local feeding density, this reduces the variance in the quality of the utilised patches, V. The variance therefore becomes a decreasing function of the *total search activity* by the foragers in the population, $\widetilde{S}$, which is the product NS of the population density N and the average individual search effort S.

The equilibrium distribution of search activity in a population, if it exists, must satisfy therefore the following conditions:

A. It must be the optimal search effort for all the individuals in the population at the equilibrium distribution of quality in the patches, i.e. on the $S^*(V)$ function.

B. The equilibrium distribution of the quality in the patches is that which is generated by the total effect of the equilibrium distribution of search effort in the population, i.e. on the $V(\widetilde{S})$ function.

C. Thus, it must be at the intersection between the $S^*(V)$ function and the $V(\widetilde{S})$ function (see Fig. 2).

D. The equilibrium distribution and variance of quality in the exploited patches is expected to provide for each individual an average marginal benefit per unit of search effort which is exactly balanced by an equal average marginal cost per unit of search effort.

E. The equilibrium is expected to be stable in most cases: (i) because both V and S are expected to change in the direction of their respective equilibrium functions; (ii) because the time constants of the two processes are usually very different. Usually the quality in large patches changes much more slowly than the behavioral adaptive response of the searching and foraging animals, but the opposite is also possible if the patches are very small.

   An oscillatory approach to the equilibrium or some unstable periodic changes are possible if the time constants of the two processes are similar. This is unlikely however because the behaviour of the animals is usually not synchronised. I did not investigate this possibility.

F. The variance of the equilibrium distribution of the quality of the patches may be quite high. This represents a distinct departure from the zero variance predicted by the IFD when there is complete information.

This equilibrium variance of the patch quality is analogous to the equilibrium price dispersion in competitive economic markets with varying inputs and incomplete information (Burdett and Judd 1983). See Figure 2.

IV. Searching and foraging in heterogeneous populations

The introduction of a very low density of type i animals which utilise the same resources, but have a different optimal $S_i^*(V)$ function from the majority, which have the function $S^*(V)$, has a negligible effect on the distribution and the variance of quality in the

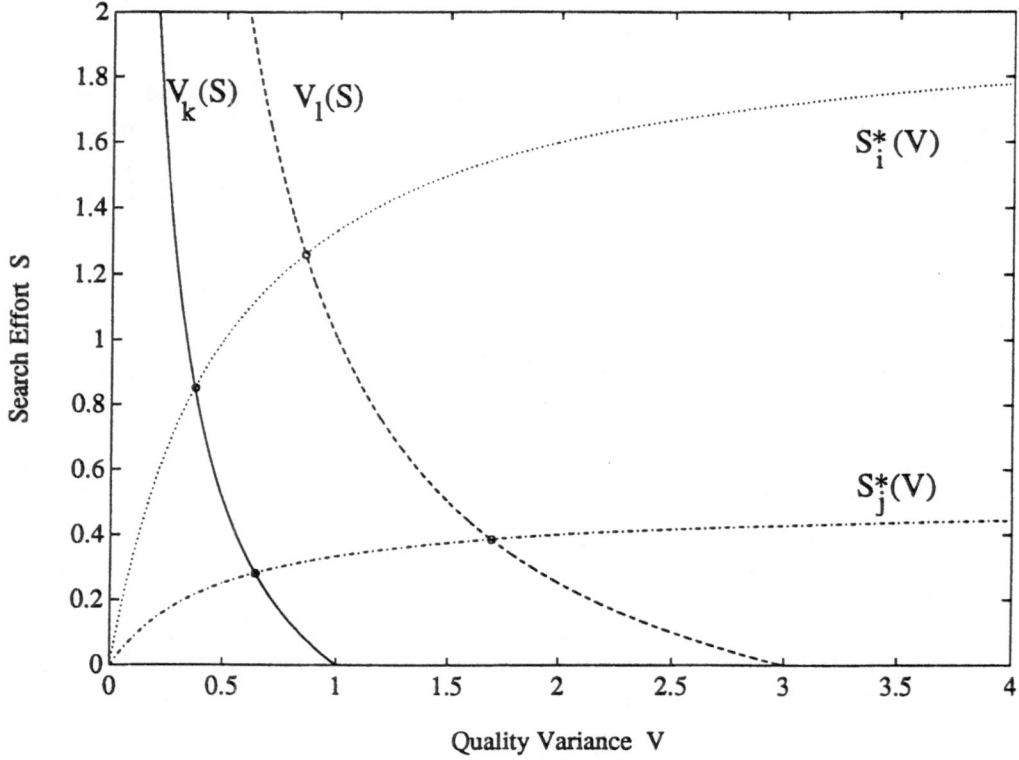

Figure 2. A schematic graphic illustration of four possible stable joint equilibrium points at the intersections between the optimal search effort functions of the variance of the patches, $S^*(V)$, of two different types i and j, and two functions that characterise the effect of the total search effort on the variance, $V(\widetilde{S})$, in two different environmental conditions k and l, in a uniform population. (Arbitrary units).

Equilibrium search effort S^{**} is highest for the high $S_i^*(V)$ and $V_\ell(S)$ functions, while the equilibrium variance V^{**} is highest with the high $V_\ell(\widetilde{S})$ function and the low $S_j^*(V)$ function. S^{**} and V^{**} may be positively or negatively correlated, or not correlated at all, depending on the distribution of different types in different environmental conditions.

patches. The equilibrium level S_i^{**} will therefore be at the optimal search effort of the i-th type at the equilibrium distribution and variance generated by the common type.

See Fig. 3 for a schematic illustration of the equilibrium of search efforts and the variance in heterogeneous and mixed populations.

If the optimal search function $S_i^*(V)$ of the rare type is higher than $S^*(V)$ of the common type, then the equilibrium level of the rare type, S_i^{**}, will be higher than the equilibrium it would reach if it were the common type alone. The opposite result is obtained if the optimal search function of the rare type is lower than that of the common type. Thus, the equilibrium search levels, S_i^{**}, of the rare types are driven to more extreme levels relative to the common type (Fig. 3A).

In the more general case of several different populations which utilise the same resources, with different $S_i^*(V)$ functions, an equilibrium of *total search activity* will be

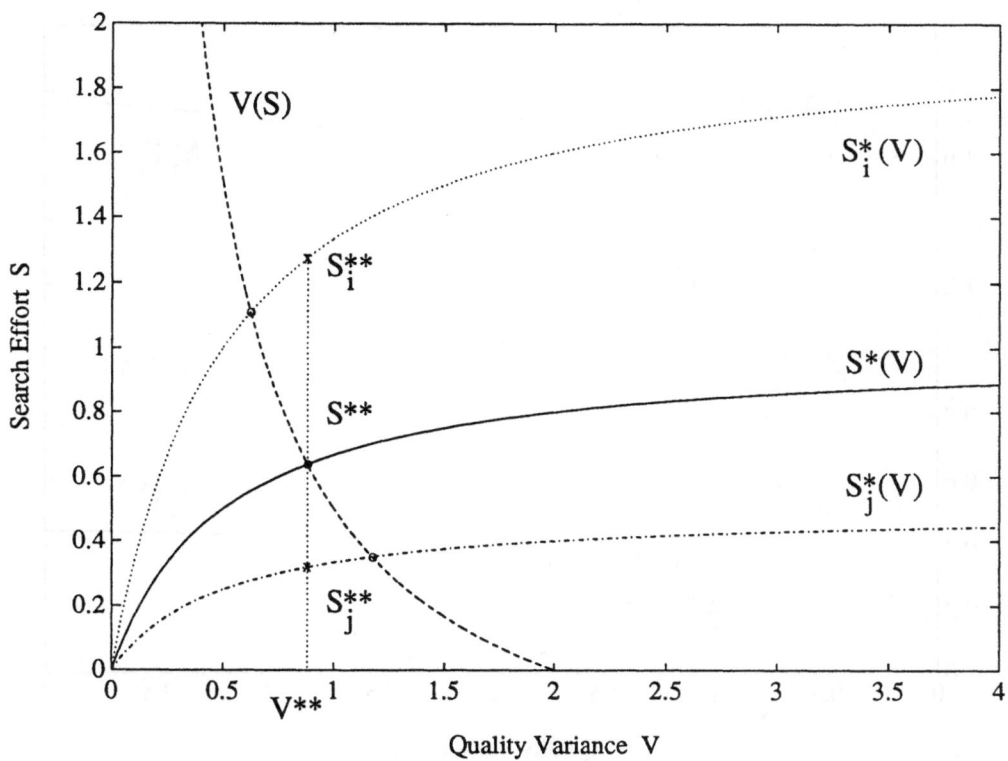

Figure 3. A schematic illustration of the joint equilibrium points between search effort and the variance of patch quality in heterogeneous populations. (Arbitrary units).
3A. One common type with $S^*(V)$ function, and two rare types, i and j, with $S_i^*(V) > S^*(V)$, and $S_j^*(V) < S^*(V)$ functions respectively. The equilibrium variance V^{**} is determined solely by the common type $S^*(V)$ function. The equilibrium points of the two rare types are each at its own optimum level as a function of this variance. As indicated in the figure, S_i^{**} is above and S_j^{**} is below their respective equilibrium points they would have if they were the common types alone, thus having more specialised extreme optimal behaviour in the presence of the common type.

reached at some intermediate equilibrium level of V^*, at which each one of the different types will be at its *own optimal $S_i^{**}(V^*)$ level*. At this joint equilibrium, the equilibrium variance V^* will be the result of the total effect of all the $S_i^{**}(V^*)$ search efforts weighted by their relative densities, $\widetilde{S}^{**}$. The range of equilibrium levels of the different types will be more widely dispersed than the range of equilibrium levels if each of them was a common type on its own (Fig. 3B).

The joint ESS search effort of foraging young and adult animals. This is an interesting special case of a heterogenous population of animals with different optimal search functions. Over an evolutionary time scale, young animals are expected to have evolved their optimal search efforts as adaptive responses to their expected cost and benefit functions.

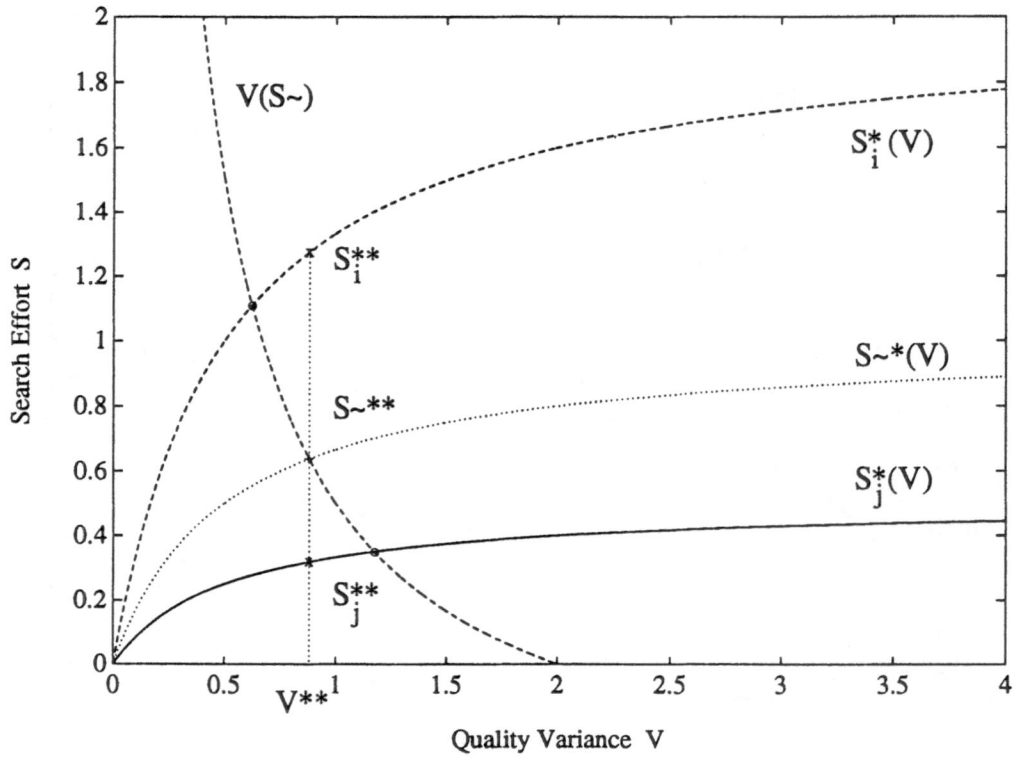

Figure 3B. Two equally common types, i and j, with different $S^*(V)$ functions. The equilibrium variance V^{**} and the equilibrium total search effort $\widetilde{S}^{**}$ are at the intersection between the $V(\widetilde{S})$ function and the optimal total search effort function $\widetilde{S}^*(V)$, which is the weighted sum of $S_i^*(V)$ and $S_j^*(V)$. The equilibrium search effort of each type is at its own optimum as a function of the jointly determined equilibrium variance V^{**}. As indicated in the figure, S_i^{**} is above and S_j^{**} is below their respective equilibrium points they would have if they were the common types alone, demonstrating a more specialised extreme optimal behaviour in the presence of the other competing type.

In general, young animals have less information about the spatial and temporal distribution of patch quality in the habitat, compared with experienced adult animals. The young animals tend therefore to forage initially in relatively low quality patches, before they had the time to search and find the good patches and resources.

The cost function of the search effort is relatively low therefore for the young animals, which increases their optimal search effort. Furthermore, the benefit function is relatively high for the young animals because they feed in lower quality patches, so that they can gain more by allocating more time to searching. Both effects increase the ESS search effort of the young animals relative to that of the adult animals.

On the other hand, the ESS search effort of the adults is expected to be relatively low for three reasons:

One obvious reason is that the optimal search effort at the steady state may be low in a slowly changing environment even without the competition by the young animals.

The second reason is the strong competition by the young animals for the discovery and exploitation of new patches. Since the young animals are forced by their ignorance to invest much time and effort in searching, they are more likely to be the first to find new good patches. This considerably reduces the benefit function for the searching adults.

The third reason is the shorter expected life span of the adult animals. The benefit of switching to a new good patch has to take into account the expected duration of the improved feeding conditions. Older animals may only have a short expected remaining life span, which reduces the overall benefit of searching for a better patch.

The expected result of these different benefit and cost functions for search effort in young and old animals is a joint ESS of the two subpopulations, with a strong differentiation between high searching-level young animals and low searching-level adults. This case corresponds to the scheme illustrated in Figure 3B.

An analytical solution of the ESS search effort. This is fairly straightforward in monotypic populations, if the optimal search effort as a function of the variance, $S^*(V)$, and the variance as a function of the total search effort, $V(\widetilde{S})$, are explicitly specified.

As a simple example, take the case of a linear $S^*(V) = kV$, and a linear $V(\widetilde{S}) = V_0(1 - bN\widetilde{S})$, where k and b are coefficients that represent the average effects of V on S^*, and of $\widetilde{S}$ on V respectively. In this case: $V^* = V_0/(1 + bNkV_0)$, and $S^* = kV^* = kV_0/(1 - bNkV_0)$.

In this simple example, the equilibrium variance is an increasing function of the initial variance, and a decreasing function of the population density, N, and of both k and b. S^* on the other hand is an increasing function of its own coefficient, k, and of the initial variance V_0, and a decreasing function of the population density N and the $V(\widetilde{S})$ coefficient b. More realistic functions are expected to give qualitatively similar results.

In general, $S^*(V)$ can be derived from any explicit benefit and cost functions of the search effort at any steady state distribution of patch quality. Such functions are extremely difficult to define, however, on the basis of known empirical parameters. The best that can be usually done is to identify certain qualitative effects of some parameters, and to derive qualitative models for their effects on the optimal search effort.

The $V(\widetilde{S})$ function also depends in very complex ways on the dynamics of patch renewal, and on the search effort and activity, density, mobility, and sensory characteristics of the animals. In this case too, it is possible only to suggest qualitative effects of some of the parameters on $V(\widetilde{S})$, and to derive the expected effects on the equilibrium of search activity in the population.

The analytical solution of the equilibrium points becomes much more difficult even with very few different types of animals. Schematic graphical solutions for the cases of two rare types with one common type, and of two common types are illustrated in Fig. 3.

V. Stochastic generation of variance by small numbers of animals or visits per patch

When the mean number of animals which visit a patch at any one time is small, e.g. less than one, and the effect of each visit on the resource level is large, the local density of animals in the patches varies in discrete increments from zero to one, two, three, etc., with a high variance, which generates a corresponding variance in the quality of the patches. The stochastically generated variance in patch quality decreases when the number of animals per patch increases, or the effect of each visit decreases.

The stochastically generated variance may be quite high if the number of animals is considerably smaller than the number of patches or sites. For example, the variance is equal to the mean in a Poisson distribution generated by rare visits with equal probability per patch. The equilibrium levels of the search effort and the variance may increase therefore considerably above the levels expected without this stochastic component.

The stochastic self-generated searching activity is expected to dominate the short term levels of searching activity in many natural situations when the number of foraging animals is small relative to the number of many small uniformly renewed patches.

Such variance in the quality of patches will be generated even in initially uniform patches. Note that in this special case, the self-generated variance will be the only source of variance that provides the reward for searching effort.

For example, this would be the typical situation in many flower-bees systems, with continuous nectar secretion by flowers. In this case, the nectar levels in a patch or inflorescence are continuously renewed, but the level of resources in a small patch or inflorescence drops temporarily following a foraging visit by a bee, and recovers during the periods between visits. For independent foraging visits and constant nectar secretion, the distribution of nectar levels in the flowers is exponential (Possingham 1988).

The variance generated by the searching activity itself is much larger than the inherent variation of the nectar renewal rates in the flowers, which change much more slowly.

A selective advantage for searching and sampling by bees will be maintained therefore at the equilibrium variance of the self generated variance of reward even in initially uniform patches or inflorescences. Such an equilibrium search effort of bees has been modeled by Motro and Shmida (Pers. Comm.).

VI. Discussion

Several interesting and possibly important consequences follow from these models:

1. The models resolve the apparent contradiction between the predicted theoretical requirement for an Ideal Free Distribution (IFD) of equal quality or fitness at all the occupied or exploited sites at the equilibrium habitat distribution of mobile animals (e.g. Fretwell 1972), and the commonly observed large variation in quality between patches. According to our models, the joint equilibrium between search effort and the variance in patch quality is the balance between the variance generating processes that increase the

variance, and the searching, sampling and movement to better patches by the foragers, which reduce the variance.

These models stress the important role of sampling and learning by young, naive, or newly arrived animals in determining the steady state distribution of search effort and patch quality. A higher turnover of the population, which results in a larger fraction of inexperienced animals, increases the total search effort in the population. However, since it takes time to find and move to the better patches, the steady state variance in patch quality at the steady state is higher than with more experienced animals.

Thus, a higher population density, or a larger steady state fraction of experienced animals, or a higher search effort, or faster learning, reduce the variance more strongly, and are expected to cause a lower variance at the steady state.

Since the optimal search effort is an increasing function of the patch quality variance, all the processes or factors which increase the variance increase the search effort at equilibrium. Similarly, processes which decrease the variance decrease the eqilibrium search effort.

Our results are analogous to those of Stephens (1987), which provided an explicit analytical expression for the optimal sampling rate in a specific model of a changing environment. Our models provide a broader qualitative framework for dealing with a wide range of such problems.

The models are analogous to search, exploration and learning in economic activities with incomplete information. The stady state characteristics and the predicitons of our models are similar to models of steady state variation of prices or salaries (Burdet and Judd, 1983; McKenna 1985).

Several conclusions follow from this model:

1.1 Since the variance of patch quality is a decreasing function of the total search effort in the population, the variance is expected to be a decreasing function of the population density of the foragers for any given individual search effort. We expect therefore that both the variance between patches and the individual search effort will decrease as population density per patch increases. This can be caused by an increased density per se, or by a decrease in the number of patches, for example by a change from many small patches to few large patches.

1.2 If the cost of searching is high, e.g. if the distance between the patches is large, the ESS level of searching will be small, and the variance in quality in the patches will remain almost at its initial intrinsic level.
    If the searching cost decreases for a given inherent initial variance of quality in the patches, the ESS level of searching increases, and this causes the variance in the quality of the patches to decrease. In this case, a *negative* correlation is expected between the search effort and the variance of quality in the patches. A negative correlation is also expected between search cost and search effort.

1.3 On the other hand, if the intrinsic variance in patch quality increases and the cost function remains constant, the optimal level of searching is expected to increase. In this case, a *positive* correlation is expected between the search effort and the variance at the ESS.

1.4 The variance in quality in the patches may therefore be either positively correlated with the level of search if the searching cost remains constant, or be negatively correlated with the level of searching activity if the inherent variance remains constant. Since both cases may occur naturally, it is necessary to have more detailed information about the system to be able to predict the expected sign of this correlation.

2. Our models can make specific testable predictions about the effects of environmental and population variables on the search effort and the quality distribution. For example:

2.1 Both quality variance and total and individual search effort are expected to increase at the equilibrium when the renewal rate or the turnover of the patches increase. Thus, a higher search effort is expected during periods of rapid or large changes in the distribution of patches or patch quality. For example, a higher search effort is expected in foraging bees in the early hours of the morning every day, because there are often large changes from day to day in flower availability. It is also possible that some bees do not remember well their experience during the previous day, and have to refresh their memories by searching, sampling and relearning the distribution of the flowers in their foraging area.

2.2 Search effort of naive newborn or newcomer animals is expected to be much larger than that of older experienced animals. The presence of older experienced animals is expected to increase the search effort of the young animals, while the presence of younger naive animals is expected to decrease the search effort of the older animals. The search effort of young animals is expected to decline with time as they sample more patches and learn more about the distribution of food in the patches. The search effort of older animals is also expected to decline very strongly as they approach the end of their life, because this reduces the remaining expected lifetime benefit of searching and finding better patches.

2.3 Both the variance of patch quality and the individual search effort are expected to decrease when the population density or the foraging effort in the population increase. Similar effects are expected when the population densities or foraging effort increase in other species which compete for the same food sources.

2.4 Search effort is expected to decrease and the quality variance is expected to increase when the search cost, e.g. distance between patches, or the energy cost of movement between patches, increase.

2.5 New patches or resources are expected to be visited at an increasing rate after they appear, and to be underutilised during the time it takes the foragers to find them and reach a steady state rate of exploitation. The rate of equilibration is expected to be faster in high density populations, in species with lower search cost, and when there is a high proportion of young animals. The equilibration is expected to be slow if the search cost is high, e.g. the distances between the new and old patches are large.

2.6 Old patches or resources, on the other hand, are expected to be overutilised to some extent and to have a below average quality, because it takes some time for the foragers in these patches to search, find and switch to better patches. The foraging intensity and the patch quality will continuously lag behind in these depleting patches.

The equilibration will be faster if the optimal search effort is higher. The optimal search effort of any one individual is expected to increase as the relative quality in its patch decreases. This will increase the probability of leaving depleted patches, and accelerate the equilibration of the utilisation in all the patches.

3. The model shows us that information is analogous to some extent to other limiting consumable resource for which foraging animals may compete. The contribution of information to fitness is indirect however. The value of information can be measured by the increase in acquisition of direct resources, such as food, which results from the optimal utilisation of the information for making better foraging decisions.

The utilisation of the information gathered by the searching animals is mediated by the directed movement of the animals from poor depleted patches into rich new patches. This leads to a gradual depletion of the newly discovered patches by the increasingly more intensive foraging in the new patches. The value of information about new food sources decreases therefore as more animals discover them because of the greater competition for the resources. In this sense, there is a direct competition for information, *which is "consumed" by its utilisation*, which represents an indirect competition for the new food patches. The better competitors for information are those animals or animal species which invest more effort and are also more efficient in gathering the information about the new food patches, and exploit the rich levels of these new relatively unexploited food sources.

4. The model for the equilibrium search effort in a heterogeneous population can be applied to a more general modelling of the ecology and evolution of search effort in foraging communities.

Different animal species can have different benefit and cost functions for search effort, which leads to different levels of optimal search efforts, and rates of information gathering. It is possible therefore to classify animals according to their specialisations for different levels of search effort and information gathering. In general, a higher rate of information gathering is expected to lead to the utilisation of richer patches and a higher rate of food intake. It is reasonable to assume however that the higher search effort also requires a higher energy expenditure. This may allow a stable coexistence of high information and high energy species, which use the high levels of the food resources, together with low information and low energy species, which utilise the low levels of the food sources left by the high energy species.

These specialised information and energy roles in pollen and nectar collecting species of solitary bees have been pointed out for several different flowers-bees communities in Israel, and in the Alps (Cohen and Shmida 1992). The larger fast-flying bee species appear to invest relatively more time searching and exploring, and are usually the first to find and exploit new flower sources. The smaller slow-flying bees appear to invest much less time on flying large distances between patches, and tend to stay longer in the same patch. Presumably, the larger bee species which have high energy requirements are able to coexist with the smaller low energy bee species, which feed on the same flowers, by finding and arriving first on the newly flowering species, and by exploiting the initially rich levels of nectar and pollen in these flowers.

Analogous interactions in communities of granivore desert rodents have been suggested by Kotler and Brown (1988). The larger faster species have been shown to spend more time and effort travelling larger distances in their search for richer food patches, relative to the smaller slower species which tend to move and search less, and to exploit poorer food patches in a more limited area.

5. A large amount of variance is generated by the foraging and exploring activities of each animal, of other animals of the same type or species, and by animals of other species. If many animals are searching and foraging without communicating and exchanging information, this self-generated variance has the same effect as other environmentally generated sources of variance. Communication between foraging animals can lead to more structured and coordinated patterns of searching and foraging. Such complex patterns are typical of socially foraging animals: their analysis is beyond the scope of this paper.

Acknowledgement

Supported by US-Israel Binational Science Foundation Grant 89/00130, and by Volkswagen Grant I/636-91 to the Ecoratio Project.

References

Burdett, J.W., and K.L. Judd. 1983. Equilibrium price dispersion, *Econometrica* 51:955–969.

Caraco, T. 1987. Foraging games in a random environment. In *Foraging Behaviour*. (eds. Kamil, A.C., Krebs, J.R., Pulliam, H.R.), pp. 389–414. Plenum Press.

Charnov, E.L. 1976. Optimal foraging: The marginal value theorem, *Theoret. Pop. Biol.* 9:129–136.

Clark, C.W. and M. Mangel. 1984. Foraging and flocking strategies: information in an uncertain environment, *Amer. Natur.* 123:626–641.

Cohen, D. and A. Shmida. 1992. The evolution of flower display and reward, *Evolutionary Biology*. (In Press).

Cook, R.M. and S.F. Hubbard. 1977. Adaptive searching strategies in insect parasites, *J. Anim. Ecol.* 46:115–126.

Fretwell, S.D. 1972. The theory of habitat distribution. In *Populations in a Seasonal Environment*. Princeton University Press. pp. 79–114

Gould, J.L. 1984. The natural history of honey bee learning. In *The Biology of Learning*. (eds. Marler, P. and H. Terrace), Springer Verlag. pp. 149–180.

Green, R.F. 1984. Stopping rules for optimal foragers, *Amer. Nat.* 123:30–43.

Green, R.F. 1987. Stochastic models of optimal foraging. In *Foraging Behaviour*. (eds. Kamil, A.C., Krebs, J.R., Pulliam, H.R.), pp. 273–302. Plenum Press.

Houston, A.I., A. Kacelnik, and J. McNamara. 1982. Some learning rules for acquiring information. In *Functional Ontogeny*. (ed. McFarland, D.J.) Pitman, Boston. pp. 140–191.

Kacelnik, A., and J.R. Krebs. 1985. Learning about food distribution. In *Behavioural Ecology: Ecological Consequences of Adaptive Behaviour*. (eds. Sibly, R.M. and R. Smith.) Blackwell, Oxford.

Kamil, A.C., J.R. Krebs, H.R. Pulliam, (eds.). 1987. *Foraging Behaviour*. Plenum Press.

Kotler, B.P. and J.S. Brown. 1988. Environmental Heterogeneity and the Coexistence of Desert Rodents, *Ann. Rev. Ecol. Syst.* 19:281–307.

Krebs, J.R. and N.B. Davies, (eds.). 1984. *Behavioural Ecology: an Evolutionary Approach*. 2nd ed. Blackwell Publications.

Krebs, J.R., and A.J. Inman. 1992. Learning and foraging: individuals, groups, and populations, *Amer. Nat.* 140:S63–S84.

Lima, S.L. 1984 Downy woodpecker foraging behaviour: efficient sampling in simple stochastic environments, *Ecology* 65:166–174.

McKenna, C.J. 1985. *Uncertainty and the Labour Market*. St. Martin's.

McNamara, J. and A.I. Houston. 1985. A simple model of information use in the exploitation of patchily distributed food, *Anim. Behav.* 33:553–560.

Mangel, M. 1985. Search models in fisheries and agriculture. *Lecture Notes in Biomathematics* 61:105–138.

Mangel, M. 1990. Dynamic information in uncertain and changing worlds, *J. Theoret. Biol.* 146:317–332.

Mangel, M. and C.W. Clark. 1986. Search theory in natural resource modelling. *Natural Resources Modeling* 1:1–154.

McCall, J.J. 1970. Economics of information and job search, *Quart. J. Econ.* 84: 113–126.

Menzel, R. 1985. Learning in honey bees in an ecological and behavioural context, *Fortschritte der Zoologie* 31:55–74

Possingham, H.P. 1988. A model of resource renewal and depletion: applications to the distribution and abundance of nectar in flowers, *Theoret. Pop. Biol.* 33:138–160.

Pyke, G.H. 1981. Optimal foraging in hummingbirds: Rule of movement between inflorescences, *Anim. Behav.* 29: 889–896.

Pyke, G.H. 1984. Optimal foraging: a critical review, *Ann. Rev. Ecol. Syst.* 15:523–575.

Real, L. 1983. Microbehaviour and microstructure in pollinator-plant interactions. In *Pollination Biology* (ed. L. Real). Academic Press. pp. 287–304.

Real, L. 1992. Information processing and the evolutionary ecology of cognitive architecture, *Amer. Nat.* 140:S108–145.

Stephens, D.W. 1987 On economically tracking a fluctuating environment, *Theoret. Pop. Biol.* 32:15–25.

Tamm, S. 1987. Tracking varying environments: sampling by hummingbirds. *Animal Behaviour* 35:1725–1734.

Waage, J.K. 1979. Foraging for patchily distributed hosts by the parasitoid *Nemeritis canescens. Jour. Anim. Ecol.* 48:353–371.

Waddington, K.D. 1983. Foraging behaviour of pollinators. In *Pollination Biology* (ed. L. Real), Academic Press. pp. 213–239.

Appendix: The benefit function of search effort

Let us assume a stationary independent cumulative probability distribution of patch quality, X, $F_X(X') = \Pr(X < X')$.

The benefit of any given search effort S can be defined by the mean benefit of sampling a coresponding number of patches N, where $N = S/e$, with $e =$ the average sampling effort per patch. The benefit of sampling a particular set of N patches is defined by $Z_N =$ the highest quality encountered during this sampling.

Sampling a random set of N patches generates a probability distribution of the highest quality, Z, which is given by:

$$F_Z(Z')_N = \Pr(Z < Z')_N = (F_X(Z'))^N \qquad (A.1)$$

In order to derive the mean best site, we obtain first the probability density function $f_Z(Z)$ by differentiating $F_Z(Z)$:

$$f_Z(Z) = dF_Z(Z)/dZ = f_X(Z) N F_X(Z)^{N-1} \qquad (A.2)$$

We then integrate and get:

$$E(Z)_N = \int_0^\infty Z f_Z(Z) dZ = \int_0^\infty Z f_X(Z) N F_X(Z)^{N-1} dZ \qquad (A.3)$$

The mean best site $(E(Z)_N$ is an increasing function of N with a decreasing first derivative.

The mean net benefit of sampling from a patch with quality X_i is the gain in fitness above X_i, i.e.:

$$B(N) = E(Z)_N - X_i \qquad (A.4)$$

from which it is possible to derive the optimal search effort for any given cost function and quality of the current patch.

If optimal search effort cannot be changed according to the current patch quality, the mean benefit of N sampling visits has to be averaged over the overall probability density distribution of both X and X_i.

As a simple example, let us take an exponential probability density function of patch quality:

$$f(X) = L e^{-LX} \qquad (A.5)$$

where $L = 1/E(X)$. In this case,

$$F_Z(Z)_N = (1 - e^{-LZ})^N \qquad (A.6)$$

and:

$$E(Z)_N = LN \int_0^\infty Z e^{-LZ} (1 - e^{-LZ})^{N-1} dZ \qquad (A.7)$$

To investigate the behaviour of $E(Z)_N$ we note that L may be considered a scale factor, which was therefore set to $L = 1.0$. (A.7) was then solved for increasing values of N with the kind help of Dr. Gail Wolkowicz using a Math Lab software package:

N	$E(Z)_N$	DELTA/N
1	1	
2	1.5	0.5
3	1.833	0.333
5	2.283	0.225
10	2.929	0.129
20	3.590	0.067
50	4.499	0.030

The values of $E(Z)_N$ scale by $1/L$ which is mean X.

This example clearly shows that the mean best site increases as the number of sampled sites increases, but at a decreasing return per visit.

Considering a different simple example of a uniform distribution of X between 0 and H: $f(X) = 1/H$, and $F(X) = X/H$: $F_Z(Z)_N = (Z/H)^N$, and $E(Z)_N = HN/(N+1)$, which has the same qualitative characteristics, and approaches H asymptotically as N increases.

LIST OF CONTRIBUTORS

Steven R. Beissinger
School of Forestry & Environmental Studies
Yale University, 205 Prospect St., New Haven, CT 06511
E-mail: FE563@YALEVM

Colin W. Clark
Institute of Applied Mathematics
The University of British Columbia, Vancouver, Canada V6T 1Z2
E-Mail: USERBIEC@UBCMTSG or Colin_Clark@mtsg.ubc.ca

Dan Cohen
Deptartment of Evolution, Systematics and Ecology
The Silberman Institute of Life Sciences
The Hebrew University, Jerusalem 91904, Israel
E-Mail: DANCOHEN@HUJIVMS
Present address:
Dept. of Ecology and Evolutionary Biology
Princeton University, Princeton NJ 08544
E-Mail: DANCOHEN@PUCC

James P. Gibbs
School of Forestry & Environmental Studies
Yale University, 205 Prospect St., New Haven, CT 06511

Wade N. Hazel
Department of Biological Sciences
DePauw University, Greencastle, IN 46135
E-mail: WNH@DEPAUW

Éva Kisdi
Population Biology Group, Department of Genetics
Eötvös University, Budapest
Múzeum krt. 4/A, Hungary, H-1088
E-mail: EAPD203@HUECO.UNI-WIEN.AC.AT

Jesús Alberto León
Instituto de Zoología Tropical, Facultad de Ciencias
Universidad Central de Venezuela
Aptdo. 47058, Caracas 1041-A, Venezuela

Géza Meszéna
Population Biology Group, Department of Atomic Physics
Eötvös University, Budapest
Puskin u. 5-7, Hungary, H-1088
E-mail: MESZENA@LUDENS.ELTE.HV

Steven Hecht Orzack
Department of Ecology and Evolution
The University of Chicago, Chicago, Illinois 60637

Jan G. Sevenster
Populatiebilogie, Zoölogisch Laboratorium,
Universiteit Leiden, Postbus 9516
NL 2300 RA Leiden, The Netherlands
Electronic mail SBU0JS@RULSFB.LEIDENUNIV.NL

Richard Smock
Department of Mathematics and Computer Science
DePauw University, Greencastle, IN 46135

Jacques J.M. van Alphen
Populatiebilogie, Zoölogisch Laboratorium,
Postbus 9516, NL 2300 RA Leiden, The Netherlands

Jin Yoshimura
Department of Zoology
Duke University, Box 90325, Durham, NC 27708-0325
E-mail: JINBEI@ACPUB.DUKE.EDU or JINBEI@DUKEMVS